Wild Berries of the Northwest
Alaska, Western Canada and the Northwestern States

Red-flowering currant

J. Duane Sept

Calypso Publishing

Text copyright © 2005 J. Duane Sept
Photographs copyright © 2005 by photographers credited

All rights reserved. No part of this publication may be reproduced, stored in a retrieval system, or transmitted, in any form or by any means, without prior permission of the publisher, except by a reviewer who may quote brief passages in a review.

Library and Archives Canada Cataloguing in Publication

Sept, J. Duane, 1950-
Wild berries of the Northwest : Alaska, western Canada and the northwestern states / J. Duane Sept.

Includes bibliographical references and index.
ISBN 0-9730390-8-6

1. Berries--Northwest, Pacific--Identification. I. Title.

QK110.S46 2005   581.4'64'09795   C2005-901376-1

Front Cover Photos: Red elderberry, high bush-cranberry, salmonberry, Sitka mountain-ash and salal by J. Duane Sept.
Back Cover Photos: Bog blueberry and red-osier dogwood by J. Duane Sept.
Printing and Binding: Kromar Printing Ltd., Winnipeg, MB, Canada.

Published by:
Calypso Publishing
P.O. Box 1141
Sechelt, BC Canada
V0N 3A0

Website: http://www.calypso-publishing.com

Disclaimer
The author, publisher, distributor and bookseller assume no liability for the actions of the reader. Every effort has been made to provide accurate information and appropriate notes of caution in this book. Some of the fruits and berries listed here are **toxic** and must not be ingested. If you have any doubt at all about the edibility of any fruit you see in the wild, refrain from eating it until you have confirmed its edibility with an expert. **Caution is advised.**

1016994078

Table of Contents

Introduction 5
  How to use this guide 5
Junipers 7
Oregon-grapes 10
Currants & Gooseberries 13
Strawberries 18
Hawthorns 20
Cherries 22
Saskatoon 25
Crabapple 27
Raspberries & Blackberries 29
Roses 37
Mountain-ashes 40
Dogwoods 42
Blueberries 45
Huckleberries 48
Cranberries 51
Bush-cranberries 53
Elderberries 56
Miscellaneus Fruits 59
Poisonous Fruits 73
Freezing, Canning and Drying 79
Recipes — Tried and True 80
Further Reading 89
Acknowledgements 90
   Photo Credits 90
Index 91
About the Author 95

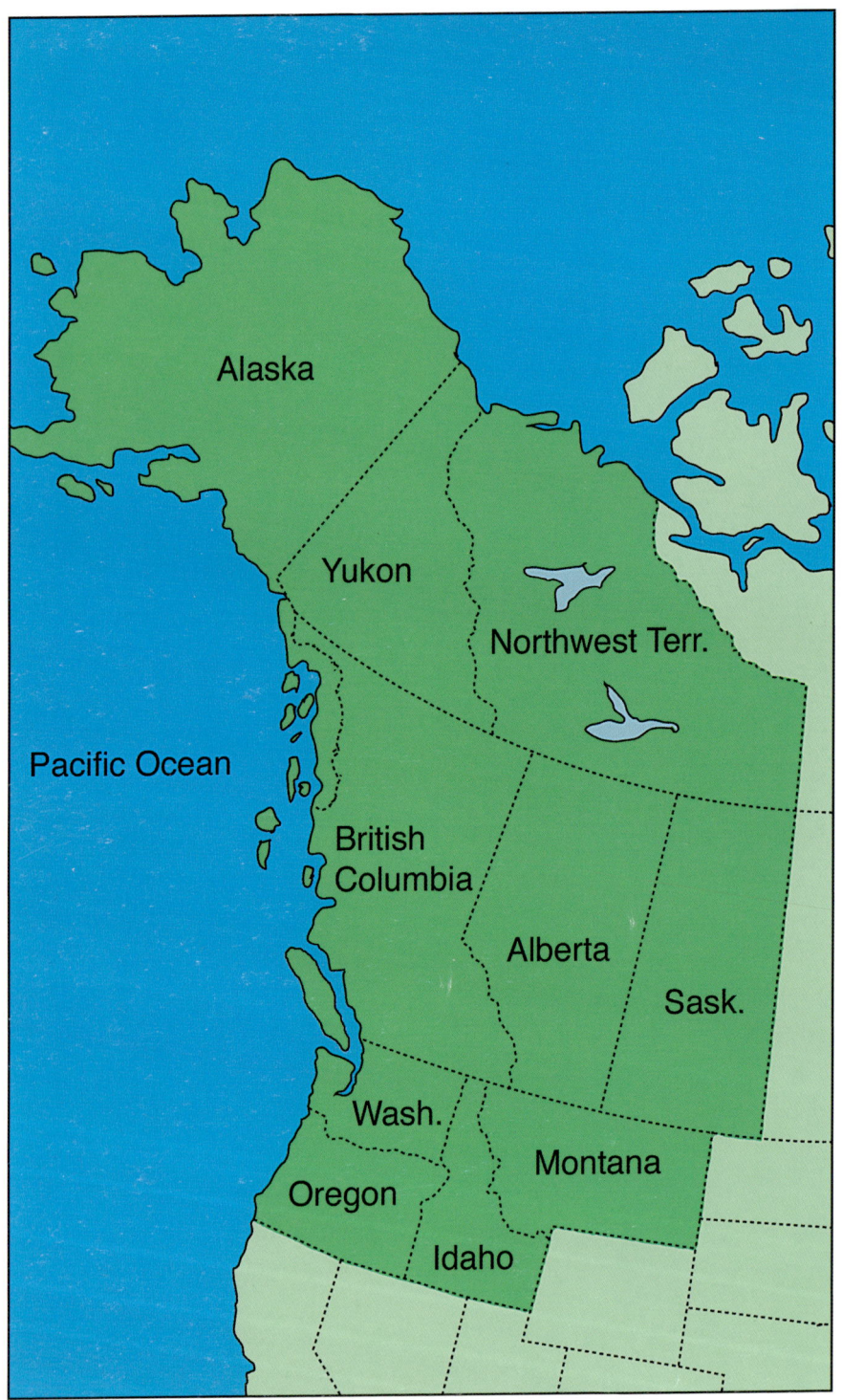

Figure 1. Geographic area covered in this guide.

Introduction

The bountiful fruits and berries of wild plants in the Pacific Northwest are a special treat. They make a perfect snack during a hike or at home in the kitchen, and they taste wonderful in jams, pies, tarts, liqueurs and other dishes. It is important to note that only some of the many abundant wild fruits that grow in our region are edible. This guide is meant to help you identify those species that are inedible or poisonous, but **exercise extreme caution** in eating any wild plant. The common species to be seen in this region are included here, but you may also come across newcomers introduced from other geographic locations, or plants that are not included in the guide.

The fruit of a plant may consist of berries, drupes, drupelets, pomes or another form. Many berries and other fruits found in this part of the world are varied and excellent tasting, but others are poisonous. Be careful.

When you observe wildlife feeding on fruit, do not assume that this fruit is edible by humans. The European holly, for example, is extremely toxic to people, but it is relished by many wild species. Harvest or taste only those species that you are absolutely sure about, or that have been identified positively by an expert.

This book includes species that occur in the areas of the Pacific Northwest highlighted in Figure 1. This large area comprises a wide variety of habitats, and a broad representation of species is covered.

Parks and preserves normally do not allow berry picking. For birds and mammals in these areas, fruits are an important source of food, crucial to their survival. As well, large numbers of berry pickers in a small park would trample the plants. Harvest wild berries in one of the many areas that are not protected by law, and that are accessible to the public.

How to Use This Guide

This guide offers a variety of information to help identify many of the common wild fruits found throughout the Northwest.

FAMILY
A plant family is a grouping of one or more genera with similar overall characteristics. All lilies, for instance, belong to the family Liliaceae, a family that includes many genera.

SPECIES NAMES
A common and a scientific name are listed for each species. Every living organism has a unique scientific name consisting of two parts: the genus (a grouping of species with common characteristics) and the species. Occasionally names change as new scientific information is discovered. The most current or appropriate name is included in this book.

Common names are those used in everyday conversation by people who live in an area where the species occur, so many plants have several common names. The most widely accepted common name appears at the top of each entry with the species' scientific name.

ADDITIONAL NAMES
Other common names and scientific names for the species are listed here.

DESCRIPTION
To identify a species, use the photograph and the written description together. Data on the flower, leaves and plant height accompany a description of the fruit. Together these features will help you identify many common species. Wildflower and fruit colors can vary significantly within a region, so use the notes on color as a general guide in the field.

HABITAT
Habitat is the type of area in which a species normally grows. Many common plants are found in more than one habitat; some plants have specific moisture requirements.

RANGE
Range is the physical location where this species is known to grow.

EDIBILITY
The fruit or berries of some species can be safely eaten by humans. These range from barely edible to delicious. Some edible species have **poisonous parts**, such as stones. Other species—fruits, leaves and/or bark—are very poisonous. Experts' opinions vary, and the information included here is the general consensus. Moderation in eating any wild species is advised. Allergic reactions can be triggered by any type of food.

Do not eat anything that you cannot identify confidently as edible. Many of the fruits and berries included in this book are toxic and must not be ingested. Look for the warning **toxic** or **poisonous** in the edibility section.

NOTES
Notes accompanying each species give special information, such as interesting features of the plant and traditional uses from eating to healing. These are intended as interesting details, not recommendations. **Caution** is advised in eating or using any wild plant.

SIMILAR SPECIES
Plants that are similar looking in appearance are identified in this section, along with notes and range.

Junipers

Junipers are famous for the distinctive taste of their blue, berry-like fruit, which is used in the making of gin. Various species of juniper have also been used for a wide range of medicinal purposes since the Middle Ages. The "berries" are actually fleshy, aromatic cones. These cones take 2 years to mature, so they are present year-round on the plant, at one stage or another. Some sixty species of junipers grow in the northern hemisphere.

Rocky mountain juniper

Cypress Family (Cupressaceae)

Common Juniper *Juniperus communis*

OTHER NAME Ground juniper.

DESCRIPTION Coniferous, evergreen shrub. **Height:** To 40" (1 m) tall. **Leaves:** Whorls of 3 needles; to ½" (1.2 cm) long. **Cones:** Male: Pollen; to 3/16" (5 mm) long. Female: Berry-like cone, dark blue; to ½" (1.2 cm) long (produced in April and May but present year-round).

HABITAT Dry, open areas and forest edges.

RANGE Alaska to New Mexico.

EDIBILITY The berry-like cones can be used as a spice in meat dishes, and to make a gin-like liqueur and juniper berry brandy. **Caution:** Some authorities consider junipers to be **poisonous**. The addition of a few berries to various dishes for flavor is considered safe, but the leaves and oil of all junipers are toxic.

NOTES In the past, junipers were used in Europe as an abortive agent, a practice that was occasionally fatal. Native people have used the species in a variety of ways, including treatment for diarrhea and heart, lung and kidney problems. Today this juniper and many others are used as ornamental shrubs in yards and gardens.

Cypress Family (Cupressaceae)

Rocky Mountain Juniper *Juniperus scopulorum*

DESCRIPTION Coniferous evergreen shrub. **Height:** To 33' (10 m) tall. **Leaves:** Opposite, scale-like in 4 vertical rows; to 1/32" (7 mm) long. **Cones:** Male: Pollen; to 3/16" (5 mm) long. Female: Berry-like cone, dark blue; to 1/4" (6 mm) long (produced in May and June).
HABITAT Open, dry rocky areas and grasslands.
RANGE British Columbia east to Alberta and Colorado, south to New Mexico.
EDIBILITY The berry-like cones can be used as a spice in various dishes. **Caution:** Some authorities consider junipers to be **poisonous**. The addition of a few berries to various dishes for flavor is considered safe, but the leaves and oil of all junipers are toxic.
NOTES This species is often thought of as a shrub in the northern part of its range, but its trunk may reach an impressive 3' (1 m) in diameter in the southern parts of its range. The plant lives as long as 350 years, and other junipers are known to have lived an unbelievable 1,500 years! Junipers have been under examination for their unique properties, including antibiotic compounds that may be useful in treating cancer.

SIMILAR SPECIES
Creeping Juniper
Juniperus horizontalis
This low-growing shrub grows to 6" (15 cm) high, with scale-like leaves that lie flat against the branch. Its edibility is similar to that of other junipers. The creeping juniper is found on dry sites from Yukon and the Northwest Territories to Wyoming.

Creeping Juniper

Oregon-grapes

Oregon-grape obtained its common name from its fruit, which strongly resembles grapes. This group of plants is noteworthy for their excellent but sour taste, and are favored in making a wide variety of preserves. The leaves strongly resemble those of European holly (see p. 76).
The fruit grows in grape-like clusters when it becomes ripe late in the summer, and is easy to gather. If it is not to be used immediately, store the fruit by freezing it.

Tall Oregon-grape

Barberry Family (Berberidaceae)

Tall Oregon-grape *Mahonia aquifolium*

OTHER NAMES Hollyleaf Oregon-grape; formerly *Berberis aquifolium*.

DESCRIPTION Evergreen shrub. **Height:** To 5' (1.5 m) tall. **Leaves:** 5–11 leaflets, very glossy above; leaflets to 3" (7.5 cm) long. **Flowers:** Yellow; to ½" (1 cm) long; in erect clusters (May to July). **Fruit:** Blue, whitish bloom or powder, grape-like berries; to ⅜" (9 mm) diameter (August and September).

HABITAT Dry, open forests.

RANGE Southern British Columbia south to northern California, west to Idaho.

EDIBILITY Edible with good taste but sour and somewhat bitter; excellent for jellies, juice, pies and wines. It adds a tangy taste to a mix of other, less flavorful berries. The fruit tastes best when it is picked fully ripe or touched with frost. **Caution:** All parts of the Oregon-grape contain berberine, an alkaloid drug that has antibiotic and analgesic properties, so consume the fruit in moderation. The roots contain the highest concentration of berberine, which is very potent and potentially **toxic**.

NOTES This handsome species, with its colorful flowers and its autumn leaves and fruit, is a great addition to gardens. It was widely used by aboriginal people, who made yellow dye from the bark and used the berries as a cure for fever.

Barberry Family (Berberidaceae)

SIMILAR SPECIES

Dull Oregon-grape *Mahonia nervosa*
This smaller species grows to 2' (60 cm) tall, with 9–19 leaflets. These leaflets have 3 main veins and a dull appearance, unlike the glossy tall Oregon-grape. Dull Oregon-grape grows along the coast from British Columbia south to California.

Dull Oregon-grape

Creeping Oregon-grape
Mahonia repens
This shrub grows to a height of 2' (60 cm) with 3–7 leaflets on each compound leaf. The teeth on the leaves are much shorter than those of tall Oregon-grape. This species favors dry sites east of the Cascades, from British Columbia south to New Mexico, including the Great Plains.

Creeping Oregon-grape

Currants & Gooseberries

The color of the fruit of currants varies considerably from species to species, ranging from green to black to bright red. All are edible, but few are tasty enough to be noteworthy. Some species have eye-catching flowers, especially the red-flowering currant and golden currant. These colorful plants are available in many nurseries and are often planted as shrubs in yards and gardens.

Currants are similar to gooseberries, but currants normally have no spines and gooseberries usually do. The name gooseberry comes from the old English custom of stuffing a roast goose with berries.

Bristly black currant

Currant Family (Grossulariaceae)

Bristly Black Currant *Ribes lacustre*

OTHER NAMES Black gooseberry, black swamp gooseberry, swamp gooseberry, prickly currant.
DESCRIPTION Deciduous shrub armed with small, yellowish prickles overall and larger spines at the leaf nodes. **Height:** To 6½′ (2 m) tall. **Leaves:** Alternate, maple leaf-shaped with 5 deeply divided lobes; to 2″ (5 cm) across. **Flowers:** Yellowish green to reddish brown, saucer-like, in drooping racemes or clusters of 7–18; to 1/8″ (3 mm) long (April to July). **Fruit:** Dark purple, round berry; to 5/16″ (8 mm) diameter (July and August).
HABITAT Moist forests, stream banks, exposed slopes; low to subalpine elevations.
RANGE Alaska to Newfoundland, south to California, Utah, Colorado, Michigan and Pennsylvania.
EDIBILITY Edible but insipid.
Caution: The spines of this species cause allergic reactions in some individuals. Caution is advised.
NOTES This species appears to be a gooseberry but is a currant that is an exception to the rule. The fruit of this species is often sparse in drier areas, whereas a single plant in a moist area may produce abundant fruit. The bristles and hairs on the berries, which have an unpleasant smell, prompted some Native tribes in British Columbia to call these berries "hairy face."

SIMILAR SPECIES

Northern Black Currant *Ribes hudsonianum*
This currant is an unarmed shrub that grows to 6½′ (2 m) tall with white flowers in erect clusters, and black fruit, which appears later in the season. It grows east of the Coast Mountains from Alaska to California, west to Utah, Wyoming and Minnesota.

Sticky Currant *Ribes viscosissimum*
This deciduous shrub reaches a height of 6½′ (2 m), with greenish white to light pink flowers. The fruit is bluish black, very sticky and not considered edible because of its disagreeable taste and smell. It can be found on dry sites from southern British Columbia to northern California, east to Arizona, Montana, Wyoming and northwest Colorado.

Sticky currant

Currant Family (Grossulariaceae)

Red-flowering Currant *Ribes sanguineum*

OTHER NAMES Winter currant, red flower currant.
DESCRIPTION Unarmed deciduous shrub. **Height:** To 10′ (3 m) tall. **Leaves:** Alternate, 5-lobed; to 2 3/8″ (6 cm) across. **Flowers:** Deep red to pink, elongated clusters of 10–20; to 3/8″ (9 mm) long (March to June depending upon elevation). **Fruit:** Bluish black with a whitish bloom, round berries, in elongated clusters; to 3/8″ (9 mm) diameter (July and August).
HABITAT Open forests, forest edges, disturbed areas; low to medium elevations.
RANGE Coastal British Columbia to California.

EDIBILITY Edible but far from incredible.
NOTES This species is well known for its fabulous floral display in the early spring, a quality that early-arriving hummingbirds surely appreciate. No other local member of the *Ribes* clan has such an impressive show of color, which is so spectacular that the species has been introduced into the gardens of Europe. Its scientific name, *sanguineum*, means "blood red." The brilliant colors of its flowers definitely make up for the berries' lack of taste.

Golden Currant *Ribes aureum*

OTHER NAME Buffalo currant.
DESCRIPTION Deciduous shrub. **Height:** To 6′ (1.8 m) tall. **Leaves:** Alternate, 3-lobed; to 2″ (5 cm) long. **Flowers:** Bright yellow, often with a red or purplish tinge, tubular, clusters of 6–15; to 1/2″ (1.3 cm) long (April and May). **Fruit:** Yellow to red or nearly black berries; to 1/4″ (6 mm) diameter (July and August).
HABITAT Lowland shrubby areas, rocky sites, wet areas.
RANGE Washington south to California, east to Saskatchewan, Colorado and Nebraska.

EDIBILITY Edible and juicy, but not highly regarded for its taste. These plants are best known for their wonderfully colorful blossoms and fruit.
NOTES The distinctive color of the flowers and berries are the key to identifying this spectacular species. A spicy aroma and spectacular floral display are its best attributes. Golden currant is available in many nurseries and is often planted in yards and gardens.

Currant Family (Grossulariaceae)

Wax Currant *Ribes cereum*

OTHER NAME Squaw currant.
DESCRIPTION Unarmed deciduous shrub.
Height: To 5' (1.5 m) tall. **Leaves:** Drab olive, in clusters at the ends of branches, 3–5 lobes; to ¾" (2 cm) long. **Flowers:** White or pink, sticky, tube-shaped, in short drooping racemes, to 5/16" (8 mm) long (March and April). **Fruit:** Red, sticky berries, clusters of 5–15 at the branch ends; to 5/16" (8 mm) diameter (August and September).
HABITAT Open dry shrubby areas and forests; low to medium elevations.
RANGE British Columbia to California, east Saskatchewan to New Mexico.
EDIBILITY The fruit of the wax currant is edible raw or cooked and can be used to make jellies, jams, sauces and pies, but its taste is not celebrated.
NOTES This currant is a resident of sagebrush and bunchgrass country. Both its scientific name, *cereum*, and its common name refer to its waxy appearance, caused by the secretions of glands on its leaves.
Aboriginal people made an infusion of the inner bark and used it medicinally as a wash for sore eyes. The fruit has been used to treat diarrhea and, eaten in quantity, to induce vomiting.

Northern Red Currant *Ribes triste*

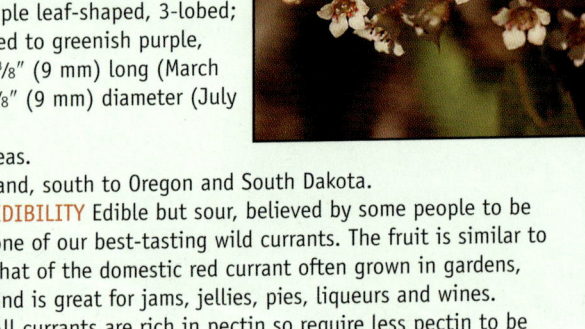

OTHER NAMES Red swamp currant, wild red currant; also known as *R. propinquum*.
DESCRIPTION Deciduous shrub, unarmed. **Height:** To 40" (1 m) tall. **Leaves:** Alternate, maple leaf-shaped, 3-lobed; to 5" (12 cm) across. **Flowers:** Red to greenish purple, drooping in clusters of 6–20; to ⅜" (9 mm) long (March to June). **Fruit:** Red berries; to ⅜" (9 mm) diameter (July and August).
HABITAT Wet forest and moist areas.
RANGE Alaska east to Newfoundland, south to Oregon and South Dakota.

EDIBILITY Edible but sour, believed by some people to be one of our best-tasting wild currants. The fruit is similar to that of the domestic red currant often grown in gardens, and is great for jams, jellies, pies, liqueurs and wines. All currants are rich in pectin so require less pectin to be added in preparation. Native Alaskan peoples added salmon roe to these berries and stored them for the winter months.
NOTES This fruit is one of the favorite currants of the Dena'ina people of Alaska. Traditionally the berries were mixed with salmon roe and stored for the winter. The fruit was also skinned and made into a poultice as a treatment for sore eyes. Native peoples also used the stems and inner bark in preparing a tea used in the treatment of colds and flu.

Currant Family (Grossulariaceae)

Northern Gooseberry *Ribes oxyacanthoides*

OTHER NAMES Canadian gooseberry, bristly wild gooseberry, smooth gooseberry, American mountain gooseberry; also known as *R. hirtellum* and *R. setosum*.
DESCRIPTION Deciduous shrub; 1–3 spines at nodes, branches bristly; to 3/8" (9 mm) long.
Height: To 5' (1.5 m) tall. **Leaves:** Alternate, maple leaf-like with 3–5 lobes; to 1½" (4 cm) across. **Flowers:** White to greenish yellow, elongated; petals to 1/16" (2 mm) long (May).
Fruit: Red to bluish purple, round berries each with a very long protuberance; to 3/8" (9 mm) diameter (July and August).
HABITAT Open woods and rocky sites.
RANGE Alaska to Newfoundland, British Columbia, Michigan, North Dakota and Montana.
EDIBILITY Edible raw or cooked with a sweet, pleasant flavor. Great in jams, jellies and other preserves.
NOTES This species is noted for its tiny flowers and abundant bristles. The leaves of the northern gooseberry turn vivid red with the first touch of frost. The traditional medicinal uses of this species by Native peoples include the treatment of bladder problems, menstrual problems and illness after childbirth. The long spines on the boughs of this shrub were used to remove splinters, probe boils and make tattoos, among other things.

Brilliant fall colors of northern gooseberry.

Strawberries

Strawberries are no strangers to anyone who frequents the outdoors. This fabulous fruit is commonly found in a wide range of habitats, from seashore to subalpine. The berries taste delicious and are often found in abundance. Their high concentration of vitamin C is an added bonus. The taste of the wild strawberry far surpasses that of the garden varieties, many of which are crosses between the American wild strawberry and a South American species.

Wild strawberry

Rose Family (Rosaceae)

Wild Strawberry *Fragaria virginiana*

OTHER NAMES Smooth wild strawberry, mountain strawberry.
DESCRIPTION Perennial herb from rhizomes, with trailing runners or stolons covered in hairs. **Size:** To 6" (15 cm) long. **Leaves:** Bluish green above, pinnate compound, terminal tooth on each leaflet smaller than its neighbouring teeth; to 4" (10 cm) long. **Flowers:** White, saucer-shaped with 5 petals, in open clusters of 2–15; to $5/16$" (8 mm) long (May to July). **Fruit:** Red; fruit stem is shorter than the leaves; to $3/8$" (9 mm) across (July and August).
HABITAT Clearings, roadsides, open forest, disturbed areas; low to subalpine elevations.
RANGE Primarily an interior species, Alaska south to California, east to the Atlantic coast and south to Colorado and Georgia.
EDIBILITY Edible and delicious, raw or cooked. Excellent in jams, preserves and pies.
NOTES Cultivated strawberries originated primarily from the wild strawberry: 90% of the parents to cultivated stock originated from this species, and the other 10% originated from the coastal strawberry (see below).
Native people used various parts of the wild strawberry plant, especially its fruit, which was usually eaten fresh but also mixed with other berries and dried into cakes for the winter.

SIMILAR SPECIES
Woodland Strawberry *Fragaria vesca*
This strawberry is identified by the terminal tooth on each leaflet, which is longer than its neighbouring teeth. Like its close relatives, this species bears excellent fruit. Woodland strawberry is found in open forests, clearings and roadsides of Europe and North America, from British Columbia to California, east to Alberta, Montana, Wyoming and New Mexico.

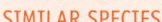

Coastal Strawberry *Fragaria chiloensis*

OTHER NAME Beach strawberry.
DESCRIPTION Perennial herb from thick rootstocks, with short trailing runners or stolons covered in hairs. **Height:** To 10" (25 cm) tall. **Leaves:** Thick and leathery, 3 leaflets; to $1\frac{1}{2}$" (4 cm) long. **Flowers:** White, saucer-shaped with 5 petals; to $1\frac{1}{4}$" (3.3 cm) across (April and May). **Fruit:** Red, strawberry; to $5/8$" (1.5 cm) across (May and June).

HABITAT Sand dunes, rocky shores and similar areas near the ocean.
RANGE Alaska to South America.
EDIBILITY Edible with fabulous taste, raw or cooked. Excellent in jams and preserves.
NOTES The coastal strawberry is easily identified by its leathery leaves and is known for its excellent taste. It often grows in rather unfavorable coastal conditions, such as in salt-spray locations, among rock fissures and on sandy beaches.

Hawthorns

The hawthorns are a group of shrubs that are easily identified, winter or summer, by the prominent spines on their branches. Authorities differ on the total number of species, from 100 to 1,200 species worldwide, and hawthorns hybridize easily. Caution is required when handling hawthorns. Scratches to the eyes can be dangerous and have even been known to cause blindness.

Black hawthorn

Rose Family (Rosaceae)

Black Hawthorn *Crataegus douglasii*

OTHER NAMES Douglas hawthorn, western thorn apple.
DESCRIPTION Deciduous shrub with a zigzag branch pattern and spines growing to 1 1/4" (3 cm) long. **Height:** To 33' (10 m) tall. **Leaves:** Alternate, leathery, oval with 5–9 lobes; to 2 3/8" (6 cm) long. **Flowers:** White, in clusters; strong-smelling; to 3/8" (9 mm) across (April and May). **Fruit:** Blackish purple pomes (apple-like fruit); to 3/8" (9 mm) long (July and August).
HABITAT Moist areas, open forest, wetland edges and similar sites.
RANGE Alaska to California, east to Saskatchewan and Michigan.

EDIBILITY Edible but contains large seeds. The fruit is used to make jellies and can be mixed with other berries to make preserves. **Caution:** The seeds contain glycoside, which has digitalis-like properties, and they are **poisonous** if eaten in large quantities.
NOTES Aboriginal people used the hard, fine-grained wood of this hawthorn in making small hand tools. The thorns were used in the fabrication of fish hooks, rakes to catch herring and lances to probe blisters.

Black hawthorn

SIMILAR SPECIES

Common Hawthorn *Crataegus monogyna*
This cultivated shrub or tree, which originated in Europe, grows to 50' (15 m) high with long, slender, curved thorns and leaves that resemble oak leaves. Its scarlet fruit is considered edible but somewhat astringent. The common hawthorn is an escapee; it occurs from southern British Columbia to California, where it is considered an invasive species.

Red Hawthorn *Crataegus columbiana*
This deciduous shrub grows to 20' (6 m) high with slender, curved thorns to 2" (5 cm) long. It produces clusters of red fruit. The red hawthorn grows from southern British Columbia to California, east of the Cascades to Idaho.

Cherries

Several species of cherries are found in the West. The flowers are small, with 5 petals and numerous stamens. All species are shrubs bearing edible fruit, and some fruit is more palatable than others. The large stone, which contains toxic hydrocyanic acid, must not be eaten with the fruit. Most wild cherries were highly regarded by Native people, who ate them fresh and dried.

Choke cherry

Rose Family (Rosaceae)

Choke Cherry *Prunus virginiana*

OTHER NAMES Chokecherry, wild cherry.
DESCRIPTION Deciduous shrub. **Height:** To 13' (4 m) tall. **Leaves:** Alternate, oval; to 4" (10 cm) long. **Flowers:** White, bottle brush-like racemes or clusters; to 6" (15 cm) long (May and June). **Fruit:** Shiny red to black; round berry-like drupes in clusters; to $^{5}/_{16}$" (8 mm) diameter (August).
HABITAT Deciduous woods, open sites and other dry areas.
RANGE Northwest Territories, Canada, to New Mexico, east to Idaho and the Great Basin.
EDIBILITY The fruit of this cherry is very astringent when not fully ripe, but sweet and delicious when the fruit ripens. In some areas, these berries have been picked after the first frost, a practice that makes them taste even sweeter. **Caution:** The pits, bark and leaves of all species of cherries (*Prunus*) contain poisonous hydrocyanic acid. There are reports that children have died from eating choke cherries without removing the **poisonous** pits.
NOTES Choke cherry is a widely available and abundant fruit that is well known to many people. Its fruit is often abundant and grows in dense clusters, making it easy to harvest, but this is offset somewhat by the time-consuming task of removing the poisonous pits. Native people harvested this cherry and added it to their pemmican, which was stored and eaten in winter when fresh food was less plentiful.

Rose Family (Rosaceae)

Pin Cherry *Prunus pennsylvanica*

OTHER NAMES Red cherry, bird cherry, fire cherry, Pennsylvania cherry.

DESCRIPTION Deciduous shrub or tree. **Height:** To 40′ (12 m) tall. **Leaves:** Alternate, lance-shaped; to 4″ (10 cm) long. **Flowers:** White, flat-topped, in clusters; to 3/8″ (9 mm) across (April and June). **Fruit:** Red, slightly elongated, in clusters; to 1/2″ (1 cm) across (July and August).

HABITAT Hillsides, riverbanks, open areas.

RANGE British Columbia to Newfoundland, south to Colorado and along the Rockies.

EDIBILITY Edible but tart, and can be eaten raw or cooked. This fruit is excellent for jellies, wines, juice, syrup and baked goods. Pin cherries contain little pectin, but they make good jams and jellies. **Caution:** The pits, bark and leaves of all species of cherries (*Prunus*) contain poisonous hydrocyanic acid, and must not be eaten.

NOTES The white floral display of the pin cherry is a spectacular show in the springtime. The fruit is eaten by a wide variety of birds and mammals, which sometimes reach the fruit before people do. In areas where bitter cherry also occurs, the two species can be distinguished by their leaves: the leaves of the pin cherry are slender with long, tapering points; bitter cherry leaves are oval without prominent points.

Bitter Cherry *Prunus emarginata*

OTHER NAME Bittercherry.

DESCRIPTION Deciduous shrub or tree with a reddish bark. **Height:** To 50′ (15 m) tall. **Leaves:** Alternate, oval; to 3″ (7.5 cm) long. **Flowers:** White, flat-topped, in clusters; to 9/32″ (7 mm) long (April and June). **Fruit:** Red to black; round; to 3/8″ (9 mm) diameter (August).

HABITAT Moist forests and stream banks.

RANGE Coastal British Columbia to California, east to Montana, Wyoming, Utah and Arizona.

EDIBILITY Very bitter fruit, as the name suggests, and generally regarded as inedible. **Caution:** The pits, bark and leaves of all species of cherries (*Prunus*) contain **poisonous** hydrocyanic acid, and must not be eaten.

NOTES This species is quite variable throughout its range with respect to size of leaves, total height and covering of soft hairs (pubescent). Several bird species consume the fruit of this cherry, including blue grouse and American robins. Chipmunks, deer mice, rabbits, deer, elk, moose and other mammals also feed on this fruit.

Saskatoon

The saskatoon is a widespread species that was traditionally very important to Native people for its abundance, taste and ease of gathering. The fruit was often dried like raisins, or dried into cakes, or mixed with dried meat and fat to make pemmican. Native people identified up to 8 different types of saskatoons based on size, texture, flowering dates and sweetness of the fruit. Scientists have also identified several species of saskatoon, although the number varies from 7 to 15 species, depending upon which authority is consulted.

Rose Family (Rosaceae)

Saskatoon *Amelanchier alnifolia*

OTHER NAMES Serviceberry, western serviceberry, Juneberry.
DESCRIPTION Deciduous shrub. **Height:** To 17' (5 m) tall. **Leaves:** Alternate, oval, toothed; to 2" (5 cm) long. **Flowers:** White, star-shaped, in small clusters; to 1" (2.5 cm) across (April to July). **Fruit:** Reddish ripening to deep purple; berry-like pomes (apple-like fruit); to 3/8" (9 mm) across (August).
HABITAT Open woods and hillsides.
RANGE Alaska to California, east to Ontario, Minnesota, Iowa, Colorado and New Mexico.
EDIBILITY Edible and sweet, especially where there is sufficient moisture. This species is renowned for its delicious taste when eaten raw or used in pies, jams, jellies, sauces and wines.
NOTES Today and in days past, saskatoons have been well known and loved for their splendid taste and abundance. They were a significant food source, especially in the making of pemmican, for many Native people. Various parts of the plant were also used as contraceptives and in the treatment of stomach troubles, childbirth recovery and other maladies. Its wood was used in making arrows, drying racks and other items.

Crabapple

The crabapple is well known because of the cultivated varieties found in gardens, but the fruit of the indigenous crabapple is much smaller. This was an important fruit to coastal aboriginal people, who gathered it and preserved it for the winter months.

Pacific crab apple

Rose Family (Rosaceae)

Pacific Crab Apple *Malus fusca*

OTHER NAMES Western crabapple, Oregon crabapple; also known as *Pyrus fusca*.
DESCRIPTION Deciduous shrub or small tree; sharp spur-shoots. **Height:** To 40′ (12 m) tall. **Leaves:** Alternate; to 4″ (10 cm) long. **Flowers:** White, very showy, fragrant, in flat-topped clusters; to ¾″ (2 cm) across (April to June). **Fruit:** Green to yellow, red when ripe, egg-shaped; to ⅝″ (1.5 cm) across (July to September).
HABITAT Coastal wetlands, moist forests; low to medium elevations.
RANGE Alaska to central California.
EDIBILITY Edible and tart when eaten raw or prepared in jellies and juices. The Pacific crab apple is high in acid content and is much sweeter when picked after the first frost. **Caution:** The seeds contain cyanide and are **poisonous** if eaten.

NOTES The Pacific crab apple is the only species of indigenous apple in western North America. The berries are high in acid and keep very well for long periods. Native people used the wood of this plant for a variety of implements, including bows, wedges and adze handles. The bark was boiled and used to treat stomach upsets, tuberculosis, ulcers and loss of appetite.

Raspberries & Blackberries

The distinctive sweet taste and bright colors of raspberries are a welcome addition to late summer. The plant that bears them is a member of the genus *Rubus*, which includes several species commonly referred to as raspberries and blackberries. Those that are "hollow" when picked are referred to as raspberries; those that are not are generally called blackberries. Red raspberries are believed to be juicier than black raspberries, and they have many fewer seeds.

Himalayan blackberry

Rose Family (Rosaceae)

Wild Red Raspberry *Rubus idaeus*

OTHER NAMES Red raspberry, American red raspberry, common red raspberry; *Rubus strigosus*.
DESCRIPTION Deciduous shrub armed with prickles that are not flattened. **Height:** To 6½' (2 m) tall. **Leaves:** Alternate, 5–7 toothed leaflets; to 4" (10 cm) long. **Flowers:** White, single to small clusters; to ½" (1 m) long (May to July). **Fruit:** Yellowish to red, hollow cluster of drupelets; to ½" (1 cm) across (July and August).
HABITAT Shrubby areas, clearings and openings.
RANGE Across Canada, east of the Cascades from British Columbia to Newfoundland, south to California and New Mexico.
EDIBILITY Edible, delectable and rich with vitamin C. Raspberries can be eaten raw or cooked, and make excellent jams, syrups, drinks, liqueurs, pies and more.
NOTES This circumpolar species is found in Europe and eastern Asia, as well as much of North America. The cultivated raspberry is very similar to the widespread wild red raspberry. Native people ate wild red raspberries fresh and mixed them with other fruit, as well as dried meat and fish, for winter food.

Rose Family (Rosaceae)

Black Raspberry *Rubus leucodermis*

OTHER NAMES Blackcap, whitebark raspberry.
DESCRIPTION Deciduous shrub with pale bluish canes, armed with curved and somewhat flattened prickles. **Size:** To 10' (3 m) long. **Leaves:** Alternate, 3-toothed leaflets; to 3" (7.5 cm) long. **Flowers:** White to pinkish, in clusters of 3–7; to 1⅛" (3 cm) across (late April to early July). **Fruit:** Red when unripe, changing to purple or black when ripe, hollow cluster of drupelets; to ½" (1 cm) across (June and August).
HABITAT Disturbed sites, forest edges, burned areas and thickets; low to medium elevations.
RANGE Alaska to California, east to Montana, Utah and Nevada.
EDIBILITY Edible and delicious. The fruit is tasty and makes good jams, jellies, preserves and pies.
NOTES The fruit of black raspberries is tasty but tends to have more seeds than wild red raspberries and to be less juicy. This species often occurs in limited numbers, but it is an excellent fruit that has been an important food source on the coast for many generations. Today, invasive Himalayan blackberry (see p. 32) often outdoes the native black raspberry for growing sites.

Rose Family (Rosaceae)

Himalayan Blackberry *Rubus discolor*

OTHER NAME Also known as *Rubus procerus*.
DESCRIPTION Deciduous shrub that begins as an erect shrub, arches and eventually begins trailing. **Size:** Sprawling to 33′ (10 m) long. **Leaves:** Alternate, 3–5 leaflets; to 10″ (25 cm) wide. **Flowers:** White to pink, 5 petals; to 1 1/8″ (3 cm) long, loose clusters (June). **Fruit:** Black, cluster of drupelets; to 5/8″ (1.5 cm) across (July and August).
HABITAT Disturbed areas, stream banks; low elevations.
RANGE East of the Cascades, from British Columbia to California.
EDIBILITY Edible and choice, raw or cooked, and excellent in jams, jellies, pies, syrups and other foods.
NOTES The Himalayan blackberry was introduced from India and has spread rapidly along the Pacific coast. It is an aggressive and invasive species that often forms an impenetrable barrier and is therefore not a good plant to introduce to the garden. It is well known, however, for producing a high volume of berries that taste absolutely fabulous.

Rose Family (Rosaceae)

Cutleaf Blackberry *Rubus laciniatus*

OTHER NAMES Evergreen blackberry, cut-leaved blackberry.
DESCRIPTION Trailing evergreen perennial armed with long, flattened prickles.
Size: To 33′ (10 m) long.
Leaves: Alternate, palmate, 3–5 deeply indented leaflets, very pronounced teeth; leaflets to 1½″ (4 cm) long.
Flowers: White to pinkish, form loose clusters; to 1⅛″ (3 cm) across (June and August). **Fruit:** Black, cluster of drupelets; to ⅞″ (2 cm) long (July and August).
HABITAT Disturbed sites, roads and waste areas.
RANGE British Columbia to California, east to Idaho.

EDIBILITY Edible, good-tasting and firmer than the Himalayan blackberry. The fruit can be eaten raw or cooked and is great in jams, jellies, preserves, pies and other foods.
NOTES Cutleaf blackberry, originally from Europe, has escaped cultivation and is establishing itself in the West. In North America it is regarded as an invasive species that is growing both in abundance and in range. This species often grows side by side with the Himalayan blackberry.

Rose Family (Rosaceae)

Trailing Blackberry *Rubus ursinus*

OTHER NAMES Dewberry, western dewberry, Douglas berry, Pacific blackberry.
DESCRIPTION Deciduous perennial, trailing to 17' (5 m) or more. **Height:** To 20" (50 cm) tall. **Leaves:** Alternate, 3 leaflets; to 2¾" (7 cm) long. **Flowers:** White or pinkish; to 1½" (4 cm) across (April to August). **Fruit:** Black, cluster of drupelets; to ⅜" (9 mm) long (July and August).
HABITAT Disturbed areas, open forest; low to medium elevations.
RANGE British Columbia to California, east to Idaho.
EDIBILITY Edible and delicious, raw or cooked. This berry is unrivaled in jams, pies and preserves.

NOTES Trailing blackberry is our only native species of blackberry and is said to be even more flavorful than the Himalayan blackberry. Male and female flowers are found on separate plants. These smaller berries make it a real treasure for the berry picker, but the plant tends to behave in an aggressive, weed-like manner and, like all blackberries, is better in the wild than in your yard or garden.

Rose Family (Rosaceae)

Salmonberry *Rubus spectabilis*

DESCRIPTION Deciduous shrub with some prickles. **Height:** To 13' (4 m) tall. **Leaves:** Alternate, pinnate, 3–5 leaflets; leaflets to 3" (7.5 cm) long. **Flowers:** Red, star-shaped; to 1½" (4 cm) across (January to June). **Fruit:** Yellow, salmon or red, raspberry-like cluster of drupelets; to 1" (2.5 cm) across (May to August).
HABITAT Moist woods, forest edges, stream banks.
RANGE Alaska to California, east to Idaho and Montana.
EDIBILITY Edible, but reviews range from "insipid" to "one of the best," likely because of a range of rainfall, temperature and other environmental factors. The fruit can be eaten raw or prepared in great jams, preserves and desserts.

NOTES People are not the only animals attracted to the wonderful fruit of the salmonberry. Many species of birds also thrive on it, including the Swainson's thrush and American robin. The salmonberry is one of the softer fruits, something to keep in mind when harvesting them. This is a tasty fruit that does not keep very long—an excellent reason to enjoy them fresh, perhaps with cream and sugar.

Thimbleberry *Rubus parviflorus*

DESCRIPTION Deciduous shrub from rhizomes. **Height:** To 10' (3 m) tall. **Leaves:** Alternate, maple leaf-shaped, 3–7 lobes; to 10" (25 cm) across. **Flowers:** White, terminal clusters of 3–11; to 1½" (4 cm) across (May to July). **Fruit:** Red, clusters of raspberry-like drupelets, hairy; to ⅝" (1.5 cm) across (July and August).
HABITAT Moist edges, clearings, open forest.
RANGE Alaska to California, east to Montana and south to New Mexico.
EDIBILITY Edible and sweet, but berries are easily crushed and fresh ones do not last long. The fruit can be eaten raw or cooked and is great in jams and preserves.
NOTES Thimbleberry flowers linger for a longer time than most and are often still present on the bush when the fruit comes in. Thimbleberries were eaten fresh by Native people, often mixed with other berries, but were not dried because the tender, juicy fruit did not store well. Other parts of the plant were also used to treat stomach aches, wounds, burns, anemia and other ailments. In China, thimbleberry has been used to treat tonsillitis, rheumatoid arthritis, hepatitis and diarrhea.

Rose Family (Rosaceae)

Five-leaf Bramble *Rubus pedatus*

OTHER NAME Creeping raspberry.
DESCRIPTION Unarmed perennial with runners, deciduous but some leaves last through the winter. **Height:** To ¾" (2 cm) tall. **Leaves:** Alternate, 5 leaflets; to 3" (7.5 cm) long. **Flowers:** White, saucer-shaped; to ¾" (2 cm) long (May to early July). **Fruit:** Red, clusters of one to several drupelets, to ⅛" (3 mm) long (August and September).

HABITAT Moist old-growth forests, stream banks.
RANGE Alaska to Oregon, east to Alberta, Montana and Idaho.
EDIBILITY Edible but small. The fruit can be eaten raw or cooked.
NOTES Five-leaf bramble often forms a mat over small sections of the forest floor. Here the plants produce their delicate flowers, which are replaced with tiny, juicy fruit that makes an excellent treat for hikers.

SIMILAR SPECIES

Dwarf Nagoonberry *Rubus arcticus*
Also known as dwarf raspberry, arctic raspberry and *Rubus acaulis*, dwarf nagoonberry grows as tall as 4" (10 cm) with pink to reddish flowers. The plant produces a small, raspberry-like fruit that is excellent for jams and jellies as well as flavoring for liqueur. It occurs from Alaska to Newfoundland, south to British Columbia, Montana, Wyoming, Colorado and Minnesota.

Cloudberry *Rubus chamaemorus*
Also called baked-apple, this small perennial grows only 8" (20 cm) high. Leaves are 5-lobed without leaflets; the flowers are white and male and female flowers grow on separate plants. The fruit is raspberry-like, varying from red to yellow in color, and edible, but described as an acquired taste. This northern species is found in Alaska and across Canada from British Columbia to Newfoundland.

Dewberry *Rubus pubescens*
This unarmed species grows to 12" (30 cm) tall with leaves that have 3–5 lobes. Its flowers are white to pinkish, transforming into raspberry-like drupelets to ⅜" (9 mm) across. The edible fruit is sweet and tangy, but the berries are so small that they are a refreshing treat rather than a substantial food. Its range extends from British Columbia to Newfoundland and Labrador, south to Washington, Colorado, Iowa, Wisconsin and Indiana.

Roses

The gorgeous flowers of the wild rose are well known, and their fruits (hips) are used in many ways. The hips are rich in vitamins A, B, C, E and K; in fact, the vitamin C content of three rose hips is equivalent to that of one orange. Rose hips stay on the branches well into the cold weather and are one of the few fruits that can be harvested during the winter.

Prickly rose

Rose Family (Rosaceae)

Prickly Rose *Rosa acicularis*

OTHER NAME Prickly wild rose.
DESCRIPTION Deciduous shrub; abundant straight prickles on twigs. **Height:** To 5′ (1.5 m) tall. **Leaves:** Compound with 3–9 leaflets, double toothed; leaflets to 2″ (5 cm) long. **Flowers:** Pink, single; to 2¾″ (7 cm) across (May to July). **Fruit:** Red, football-shaped; to ⅝″ (1.5 cm) long (July to December).
HABITAT Open woods and clearings; widespread.
RANGE East of the Cascades, Alaska to Quebec, south to New Mexico; circumpolar.
EDIBILITY The outer fleshy covering of the fruit or "hip" of all wild roses is edible, but the seeds and hairs should be removed. Native peoples believed that eating the hairs would give them "itchy bottoms." It is possible that the prickles deterred aboriginal people from using the fruit as anything but a famine food. The hair and seeds can be strained out easily after cooking, or you can taste a fresh rose hip—peel off the outer covering first. A wide variety of syrups and jellies can be made from this fruit, or it can be added to other fruits to add an interesting flavor. **Caution:** Eating too many rose hips or rose petals may cause diarrhea.

NOTES The prickly rose is a widespread species over much of the northwestern portion of the continent. It is well named with distinctive prickles, and it is the official flower of the province of Alberta. Domestic rose varieties are sometimes found in unlikely "wild" locations. They can be distinguished from our native species by their curved prickles.

SIMILAR SPECIES
Common Wild Rose *Rosa woodsii*
Also called the Wood's rose, this species lacks the numerous prickles found on the twigs of the prickly rose. Despite its name, this species is not as common as the prickly rose. It is found east of the Cascades, from Alaska south to California, east from Manitoba to Texas.

Rose Family (Rosaceae)

Nootka Rose *Rosa nutkana*

DESCRIPTION Deciduous shrub; few prickles on twigs. **Height:** To 10′ (3 m) tall. **Leaves:** Compound with 5–7 leaflets, single or double toothed; leaflets to 9/32″ (7 mm) long. **Flowers:** Pink, single; to 3″ (7.5 cm) across (May and June). **Fruit:** Red hips with hairy achenes; to 3/4″ (2 cm) long (July to November).
HABITAT Open woods, shrubby areas, open meadows, shorelines and similar areas.
RANGE West of the Cascades, Alaska to northern California.
EDIBILITY Hips are edible. They are often used in teas, wine, juice, preserves and even pies. **Caution:** Eating too many rose hips or rose petals may cause diarrhea.

NOTES The flowers of the Nootka rose, like those of all members of the rose clan, can be used in a variety of ways. They can be candied or made into a scented jelly; bakers can try rose petal cupcakes; and rose petals can even be added to salads to provide an extra touch of color.

SIMILAR SPECIES

Baldhip Rose *Rosa gymnocarp*
The hips of this species, which vary in color from orange to red, are easy to identify because the fruit is produced without sepals. The baldhip rose occurs west of the continental divide, from British Columbia to California.

Clustered Wild Rose *Rosa pisocarpa*
As its common name suggests, this species produces flowers that grow in clusters at the ends of its branches. This rose is found only in swampy locations west of the Cascade summits, from British Columbia to California.

Mountain-ashes

The colorful fruit of the mountain-ash is great to look at but was not eaten by many Native people. These fruits are sometimes used in making jams and wines. In the fall, the bright-colored leaves and fruit of this plant are striking. The European mountain-ash (*Sorbus aucuparia*) is another species, introduced from Europe and available at nurseries for planting in yards and gardens.

Sitka mountain-ash

Rose Family (Rosaceae)

Sitka Mountain-ash *Sorbus sitchensis*

DESCRIPTION Deciduous shrub. **Height:** To 13′ (4 m) tall. **Leaves:** Alternate; 7–11 leaflets with rounded tips, toothed along upper half of margin; to 8″ (20 cm) long. **Flowers:** White, flat-topped terminal clusters; to 3/8″ (9 mm) long (June and July). **Fruit:** Red to orange or purple, whitish bloom (powdery covering); berry-like pomes; to 3/8″ (9 mm) diameter (July to October).
HABITAT Open coniferous woods, stream banks, parklands; medium to alpine elevations.
RANGE Alaska and southern Yukon to northern California.
EDIBILITY Edible and bitter, but sweeter after the first frost. Historically the fruit was often stewed. It can be used to make jams, jellies, pies and wines. **Caution:** The seeds of this fruit contain cyanide and are **poisonous**. Do not eat them.
NOTES This species is found throughout the Cascades and Olympic Mountains, where it attracts several species of birds that relish the fruit. Various other species of Sorbus grow in western Europe, the Himalayas, Japan and other parts of the world.

SIMILAR SPECIES
Western Mountain-ash
Sorbus scopulina
This species grows to 20′ (6 m) high with 11–13 yellowish green leaflets that have pointed tips and teeth along most of the margin. The fruit is glossy orange to scarlet. Western mountain-ash is found primarily in the interior, from Alaska south to California, east to New Mexico.

Western mountain-ash

Dogwoods

Native dogwoods are a very diverse group of plants, varying in size from bunchberry, a small perennial that grows to a height of 10" (25 cm), to Pacific dogwood, a tree that reaches 66' (20 m) in height. But both species belong in the same genus, *Cornus*, because they have similar flowers. Other species of dogwoods are found in the temperate regions of North America, Europe and Asia.

Bunchberry

Dogwood Family (Cornaceae)

Bunchberry *Cornus canadensis*

OTHER NAMES Canada dogwood, dwarf dogwood; *C. unalaschkensis*.
DESCRIPTION Perennial from a rhizome. **Height:** To 10" (25 cm) tall. **Leaves:** Terminal whorl, oval; to 3" (7.5 cm) long. **Flowers:** Greenish white to purple cluster of flowers, surrounded by 4 greenish white bracts; to 2" (5 cm) diameter (May and June). **Fruit:** Red, berry-like drupes; to 5/16" (8 mm) diameter (July and August).
HABITAT Moist woods and clearings.
RANGE Alaska to Greenland, south to California; Asia.
EDIBILITY The fruit is edible but ranges from sweet to tasteless. It is sometimes used in puddings, sauces and jams, and as a simple trail snack.
NOTES The prominent outer parts of the greenish white bunchberry flowers are actually bracts. Their flowers have a unique and very interesting technique for pollination called an explosive pollination mechanism. Upon the release of an antenna-like trigger in the flower, the four anthers snap upward simultaneously, and the petals "pop" open and release the pollen grains in a powerful burst of energy. This is a much more effective technique for dispersing the pollen, which is quite heavy, than even a strong wind could accomplish.

Dogwood Family (Cornaceae)

Red-osier Dogwood *Cornus sericea*

OTHER NAMES Red willow; *C. alba, C. stolonifera, Svida sericea*.
DESCRIPTION Deciduous shrub. **Height:** To 10' (3 m) tall. **Leaves:** Opposite, oval to lance-shaped; to 3" (7.5 cm) long. **Flowers:** White, flat-topped clusters; to 12" (5 cm) diameter (May to July). **Fruit:** White, berry-like drupes; to ¼" (6 mm) diameter (August to October).
HABITAT Moist woods, openings, forest edges.
RANGE Alaska to California, east to Montana and Nevada.
EDIBILITY Generally considered inedible, but aboriginal people ate them in the past. To soften their bitter taste, they mixed them with other more flavorful fruit. The fruit is enjoyed by bears and by birds, especially evening grosbeaks, band-tailed pigeons, American robins and wood ducks.

NOTES The brightly colored branches of the red-osier dogwood are like candy for the eyes over the winter months, as are the clusters of white fruit in the fall. Native people used this plant to make medicinal teas in treating dizziness, chest troubles, coughs, fever, pain and other ailments. The soft inner bark was dried and smoked, alone or with common bearberry (see p. 71) or tobacco.

SIMILAR SPECIES
Pacific Dogwood *Cornus nuttallii*
This impressive tree grows to 66' (20 m) tall. In spring it is covered in gorgeous white flowers, which are replaced with clusters of green to bright scarlet fruit in autumn. This fruit is not edible. The Pacific dogwood is the provincial flower of British Columbia, where it is a protected species. It can be found along the coast from British Columbia to California.

Pacific dogwood

Blueberries

Blueberries are members of the genus *Vaccinium*, which also includes species commonly called huckleberries (see p. 48). Blueberries are so called because they tend to be blue in color. They are excellent raw or cooked, in jams, jellies, pies and many other dishes. The sweet taste of blueberries is hard to beat.

Aboriginal people, especially northern groups, used blueberries as one of their prime foods. The plants were also utilized for medicinal uses that varied from treating cancer and headaches to preventing miscarriages.

Bog blueberry

Heath Family (Ericaceae)

Oval-leaf Blueberry *Vaccinium ovalifolium*

DESCRIPTION Deciduous shrub; young twigs reddish, highly branched and grooved. **Height:** To 6½' (2 m) tall. **Leaves:** Alternate, oval; to 1½" (4 cm) long. **Flowers:** Pinkish and globular, style does not project beyond mouth of corolla, flowers appear before leaves or with them; to ¼" (7 mm) long (April to July). **Fruit:** Bluish black berries with a bluish bloom (powdery covering); to ⅜" (9 mm) diameter (June and August).
HABITAT Moist coniferous forests and forest edges; low to subalpine elevations.
RANGE Alaska to Oregon, east to Quebec, Montana and Michigan.

EDIBILITY Edible and choice, eaten raw or cooked in jams, preserves and baked goods.
NOTES The oval-leaf blueberry is highly regarded as an edible fruit. In the past, Native people ate the berries raw and also dried them into cakes to preserve them. This species often grows in proximity to the Alaskan blueberry (see below).

Alaskan Blueberry *Vaccinium alaskaense*

DESCRIPTION Deciduous shrub; young twigs yellowish green. **Height:** To 6½' (2 m) tall. **Leaves:** Alternate, oval, tiny hairs on underside of midvein; to 2½" (6 cm) long. **Flowers:** Bronze, pink or green, bell-shaped, normally broader than high, style extends beyond the flower, flowers usually appear with emerging leaves; to ¼" (7 mm) long (May and June). **Fruit:** Bluish black to purplish black, round berries; to ⅜" (9 mm) diameter (July and August).
HABITAT Moist coniferous forests, forest edges, clearings; low to subalpine elevations.
RANGE Alaska to Oregon.
EDIBILITY Edible with good flavor. Can be eaten raw or cooked and is great in jams and preserves. These berries, like many fresh fruits, freeze well.
NOTES The Alaskan blueberry favors moist forests along the coast. It is very similar to the oval-leaf blueberry (see above), though its fruit is not as flavorful. In the past this species was so abundant that it provided both fresh and dried food to most local Native groups.

Heath Family (Ericaceae)

Bog Blueberry *Vaccinium uliginosum*

OTHER NAMES Bog bilberry, western huckleberry; *V. occidentale*.
DESCRIPTION Deciduous shrub, young branches yellowish green. **Height:** To 2' (60 cm) tall. **Leaves:** Alternate, elliptical to oval with the broadest portion above the middle, leaves vary from hairless to hairy; to 1⅛" (3 cm) long. **Flowers:** Pink, urn-shaped with 4 lobes; to ¼" (6 mm) long (May to July). **Fruit:** Blue berries covered in a fine, whitish bloom (powder); to ⅜" (9 mm) diameter (July and August).

HABITAT Low-elevation bogs, subalpine heath, alpine tundra.
RANGE Alaska to northern California.
EDIBILITY Edible, sweet and choice, eaten raw or cooked. Excellent in jams and preserves.
NOTES The tasty fruit of this species was and continues to be an important food to aboriginal peoples of Alaska, northern Canada and others living along the coast farther south. It is relished by all who taste it. The bog blueberry is found in a wide range of habitats that can modify its growth pattern. In alpine situations, heavy snow can flatten the shrub, causing it to grow in a flattened form and reach a height of only about 1" (2.5 cm) and an area exceeding 1 sq ft (.09 sq m).

SIMILAR SPECIES

Dwarf Blueberry *Vaccinium caespitosum*
This species, sometimes called dwarf bilberry, is found in low-elevation bogs as well as wet alpine and subalpine meadows, where it grows to 12" (30 cm) high. Its leaves are distinctly toothed on the upper half with a bright green color above and below. Its narrow flowers are white to pink with 5 lobes. The blue fruit is covered in a light gray bloom (powder) and many people consider it the best species of wild blueberry. It can be found growing along the Pacific coast from Alaska south to California and east to the Rocky Mountains.

Velvet-leaf Blueberry *Vaccinium myrtilloides*
This species grows to 20" (50 cm) tall with leaves that are lance-shaped and covered with soft hairs. It produces white or pinkish bell-shaped flowers, and fruits that grow alone or in clusters and that have a heavy bluish bloom (powdery covering). It grows in coniferous forests east of the Cascades, from British Columbia to Washington, east to Labrador and New Mexico in the Rocky Mountains.

Huckleberries

The tasty fruit of huckleberries, close cousins of blueberries, are favored by both humans and wildlife. Huckleberries are black, purple or red in color. Both huckleberries and blueberries are members of the genus *Vaccinium*, but blueberries tend to be sweet and huckleberries are often sweet and tart. Both species are often found in great abundance, which makes them even more desirable. Huckleberry Finn was one of the better-known admirers of these berries.

For Native peoples in the region, the fruit of this species was an important part of the diet. A related species, false huckleberry (*Menziesia ferruginea*), produces pink berry-like structures under its leaves that are actually the fruiting body of the fungus *Exobasidium vaccinii*. The Tsimshian people of coastal B.C. ate this fungus, which they believed to be the snot of Henaaksiala, a mythical being that stole corpses.

Red huckleberry

Heath Family (Ericaceae)

Black Huckleberry *Vaccinium membranaceum*

OTHER NAMES Mountain huckleberry.
DESCRIPTION Deciduous shrub. **Height:** To 5′ (1.5 m) tall. **Leaves:** Alternate, elliptical to lance-shaped, sharp-pointed; to 1½″ (4 cm) long. **Flowers:** Pink, urn-shaped; to ¼″ (6 mm) long (April to June). **Fruit:** Purplish to black berries without a bloom (powdery covering), to 5/16″ (8 mm) diameter (July to October).
HABITAT Moist to dry forest understory, burned areas, mountain slopes; low to medium elevations.
RANGE British Columbia to northern California, east to Montana and Idaho.

EDIBILITY Edible and fabulous. Can be eaten raw or cooked and is excellent in jams, preserves and pies.
NOTES The fruit of the black huckleberry is reported to be one of the best tasting of the huckleberry and blueberry clan. It was and still is used by many Native peoples in the area. The berries are an excellent hiking snack and can be gathered in quantity for preparation and preserving. This fruit is also a favorite of bears, which gorge themselves on it in the fall.

SIMILAR SPECIES
Evergreen Huckleberry *Vaccinium ovatum*
This distinctive species grows to 13′ (4 m) high, with leathery evergreen toothed leaves. Its pink bell-shaped flowers grow in clusters of 3–10. The fruit is deep purplish black berries that are edible and sweet but somewhat musky tasting. This huckleberry grows from British Columbia to California.

Heath Family (Ericaceae)

Red Huckleberry *Vaccinium parvifolium*

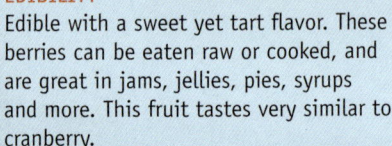

DESCRIPTION Deciduous shrub, branches bright green.
Height: To 13′ (4 m) tall. **Leaves:** Alternate, oval; to 1/8″ (3 mm) long. **Flowers:** Pink to greenish, bell- to urn-shaped; to 3/16″ (5 mm) long (April to June). **Fruit:** Red berries; to 3/8″ (9 mm) diameter (July and August).
HABITAT Coniferous forests, forest edges, often in soils with decaying wood; low to medium elevations.
RANGE Southeast Alaska to central California.

EDIBILITY
Edible with a sweet yet tart flavor. These berries can be eaten raw or cooked, and are great in jams, jellies, pies, syrups and more. This fruit tastes very similar to cranberry.
NOTES The red huckleberry is a distinctive plant with tasty fruit that is often produced for extended periods. Each berry grows and ripens quickly, but there is a steady progression of ripening fruit. This berry, like many others, is a favorite of various birds, including the American robin.

Grouseberry *Vaccinium scoparium*

OTHER NAMES Grouse whortleberry, littleleaf huckleberry.
DESCRIPTION Low deciduous shrub; branches green and strongly angled. **Height:** To 10″ (25 cm) tall. **Leaves:** Alternate, ovate; to 5/8″ (1.5 cm) long. **Flowers:** Light pink, urn-shaped; to 3/16″ (4 mm) long (May to August). **Fruit:** Bright red berries; to 3/16″ (5 mm) diameter (September).
HABITAT Open coniferous forest, alpine; medium to high elevations.
RANGE British Columbia to California.
EDIBILITY Edible, sweet and wonderful. They are so small that they are usually eaten raw unless they are very plentiful, and they can be gathered with a comb. Where abundant, they may be harvested in large enough quantities for use in jams, jellies, pies, syrups and other foods.
NOTES This fabulous-tasting fruit is particularly appreciated after a hike. It often grows in abundance in alpine areas. Its fruit is a favorite of many small mammals, including chipmunks, red squirrel, gray fox, red fox and skunks. Spruce grouse, ptarmigans, ruffed grouse, blue grouse, bluebirds, thrushes and other birds also favor these choice berries. The entire plant is also browsed upon by mountain goat, mule deer, moose and bear. It's a wonder that there are ever any berries left for humans to pick!

Cranberries

Wild cranberries are small but their tangy taste makes up for it. Although they are similar to blueberries and huckleberries, their taste is so distinctive and so much more tart that they are commonly referred to as a different berry. They are favorites at Thanksgiving and Christmas with their rich, piquant flavor, and they go well with a wide variety of other foods, including meat dishes, baked goods, juices and other drinks. Cranberries are also well known for their therapeutic properties in treating bladder and kidney ailments.
Bush-cranberries are not related to cranberries. They are described on page 53.

Bog cranberry

Heath Family (Ericaceae)

Bog Cranberry *Oxycoccus oxycoccos*

OTHER NAMES Small cranberry, wild cranberry; *Vaccinium oxycoccos, V. microcarpus, Oxycoccos microcarpus, O. quadripetalus*.
DESCRIPTION Evergreen shrub. **Height:** To 16" (40 cm) tall. **Leaves:** Alternate, evergreen, leathery, elliptical; to 3/8" (9 mm) long. **Flowers:** Pink, similar to shooting-star flowers; to 1/2" (1.2 cm) long (June and July). **Fruit:** Pink to red berries; to 3/8" (9 mm) diameter (September and October).
HABITAT Bogs, muskegs and other sphagnum environments; low to subalpine elevations.
RANGE Alaska to Labrador, south to Oregon, east to Idaho.
EDIBILITY Edible, juicy and choice, eaten raw or cooked. Delicious in jams, preserves, drinks and other foods.

NOTES The bog cranberry grows in northern environments around the globe. Aided by fungi, the plant absorbs nitrogen and phosphorus nutrients from its acidic bog environment. This species is one of the first colonizers after a fire, and after repeated fires it regenerates dramatically. In fact, fires are often set deliberately to increase the productivity of commercial cranberry operations. The plant is able to survive fires because its rhizomes reach down well below the surface of the bog.

SIMILAR SPECIES
Lingonberry *Vaccinium vitis-idaea*
Also known as lowbush cranberry or mountain cranberry, the lingonberry is related to the bog cranberry. It resembles bearberry (see p. 71). It grows to a height of 8" (20 cm) with leathery evergreen leaves with dark glands on the undersides. Its pinkish bell-like flowers are replaced with red berries. Its fruit is quite pleasant-tasting with an acidic flavor. Some think lingonberries taste better than cranberries. This species occurs from Alaska to Newfoundland, south to British Columbia.

Bush-cranberries

Bush-cranberries are not related to cranberries (see p. 51), but they have a similar tart taste. Each drupe produces a large seed, and the high acid content of these juicy berry-like drupes adds to their distinctive taste. Fall brings out the bright colors of the fruit and leaves of the bush-cranberry, and these shrubs are frequently planted in gardens where their flowers, leaves and fruit add wonderful color. The fruit also attracts a variety of birds to the backyard.

Bush-cranberries are very easy to harvest, and the stones can be removed with the skins by simmering the berries in water and letting them drain in a fine nylon jelly bag. The juice can then be made into a wonderful tangy jelly.

High bush-cranberry

Honeysuckle Family (Caprifoliaceae)

High Bush-cranberry *Viburnum edule*

OTHER NAMES Squashberry, moose berry, highbush-cranberry.
DESCRIPTION Deciduous shrub.
Height: To 12′ (3.5 m) tall.
Leaves: Opposite, 3-lobed with teeth; to 4″ (10 cm) wide.
Flowers: White, clusters of less than 50 similar flowers; to 1⅛″ (3 cm) wide (April to July). **Fruit:** Red to orange berry-like drupes, large flat stone present; to ⅝″ (1.5 cm) long (August).
HABITAT Moist forests, wetland edges; low to medium elevations.
RANGE Alaska to Newfoundland, south to northern Oregon, Idaho, Colorado, Minnesota and Pennsylvania.
EDIBILITY Edible and juicy with a high acid content. Sweeter when picked after the first frost. The stones must be removed before eating.
NOTES These tart berries sweeten on the bush over time. Native people have used them as fish bait because they resemble salmon eggs (roe), favored by several species of fish. The leaves of this plant turn to a brilliant red color in autumn. This species is called both low bush-cranberry and high bush-cranberry, which causes great confusion.

Honeysuckle Family (Caprifoliaceae)

American Bush-cranberry *Viburnum opulus*

OTHER NAMES, Americanbush cranberry, Pembina; *Viburnum trilobum* var. *americanum*.
DESCRIPTION Deciduous shrub.
Height: To 13′ (4 m) tall. **Leaves:** Opposite, 3-lobed with deeply spreading lobes; to 5″ (12.5 cm) long. **Flowers:** White, flat-topped clusters with outer flowers enlarged; inner flowers to 3/16″ (4 mm) diameter, outer flowers to 3/4″ (2 cm) diameter (May and June).
Fruit: Red to orange berry-like drupes, hanging clusters; to 1/2″ (1 cm) diameter (July and August).
HABITAT Poplar forests and moist open woods.
RANGE Central British Columbia to the Atlantic Coast, south to Washington (state).

EDIBILITY Edible but acidic and tart; sweeter once touched by frost. The berries are excellent for a wide range of jellies and other preserves, and for flavoring meat and game.

NOTES This is a common species of the eastern regions of North America whose range extends into the West. Its distinctive white flowers grow in clusters, with small flowers surrounded by larger ones. Native people ate American bush-cranberries fresh or mixed them with fat and stored them in birchbark baskets in underground caches. These delicacies were considered very valuable trading items and treasured gifts.

SIMILAR SPECIES
Oval-leaf bush-cranberry *Viburnum ellipticum*
This deciduous shrub grows to 10′ (3 m) tall with oval, hairy-stalked leaves. Its flowers grow in larger clusters to 2″ (5 cm) wide, each of which is replaced with black, oval, berry-like fruit that is edible. This species occurs from Washington to California.

Elderberries

Elderberries are easy to identify. They are rich in vitamin C, and they make great wines, sauces and jellies. The fabulous colors of elderberry fruits are another wonderful attribute. The fruit attracts an amazing variety of birds, which makes elderberries popular backyard shrubs.

Clusters of BB-sized elderberry fruit can be collected quickly, so elderberries are easy to harvest in quantities large enough to make wines, jellies and other treats. Red and black elderberries, however, should be cooked before they are eaten. As well, the leaves, bark and roots are purgative when eaten in moderate amounts.

Red elderberry

Honeysuckle Family (Caprifoliaceae)

Red Elderberry *Sambucus racemosa* ssp. *pubens* var. *arborescens*

OTHER NAMES Red elder, red-berried elder, Pacific red elder; formerly *S. melanocarpa*, *S. pubens*.
DESCRIPTION Deciduous shrub. **Height:** To 20′ (6 m) tall. **Leaves:** Opposite, pinnate with 5–7 leaflets; leaflets to 6″ (15 cm) long. **Flowers:** Creamy white, rounded cluster; to ¼″ (6 mm) long (March to July). **Fruit:** Bright red clusters; to ¼″ (6 mm) diameter (June to August).
HABITAT Stream banks, moist clearings and other spots; low to medium elevations.
RANGE Along the coastal mountain ranges, Alaska to California.
EDIBILITY The flowers are edible raw or cooked. The raw fruit is not poisonous, as is commonly believed, but it does contain hydrocyanic acid, which can cause severe intestinal upset. Native peoples often ate the flowers raw, but few people regard them highly as food. Once cooked, they can be made into a variety of preserves. **Caution:** The roots, leaves, wood and bark of red elderberry are all **poisonous**.

NOTES Elderberries are colorful shrubs that produce abundant berry-like drupes. The distinctive fruit is used to flavor a popular Italian liqueur, appropriately called Sambuca. Elderberries are available at nurseries because they make excellent ornamentals and barriers. Bears are fond of these fruits, but so are many types of birds—a bonus to yards and gardens with the elderberry's wonderful blossoms and fruit. Folklore suggests that the wood of this species can be used to relieve such ailments as toothaches, colds and freckles.

SIMILAR SPECIES
Black Elderberry *Sambucus racemosa* ssp. *pubens* var. *melanocarpa*
This shrub grows as tall as 10′ (3 m), with white flowers in round clusters that are replaced by handsome black or purplish black fruit. It grows in the Rocky Mountains from Alberta and British Columbia, south to California and Nevada.

Honeysuckle Family (Caprifoliaceae)

Blue Elderberry *Sambucus cerulea*

DESCRIPTION Deciduous shrub.
Height: To 13′ (4 m) tall.
Leaves: Opposite, pinnate with 5–9 leaflets; leaflets to 6″ (15 cm) long. **Flowers:** Cream, to ¼″ (7 mm) diameter; flat-topped clusters (May to July). **Fruit:** Powder blue, berry-like; to ¼″ (6 mm) diameter; clusters (July and August).
HABITAT Moist sites and slopes, along watercourses.
RANGE British Columbia to California, east from Montana to Arizona and New Mexico.
EDIBILITY Edible. Great in jams, juices, syrups, pies and wines. The flowers can also be dipped in batter and fried into fritters.
NOTES The scientific name *cerulea* comes from the Latin *caeruleus*, meaning "sky blue"—an appropriate name for the fruit of this plant. This species displays its flowers and fruit later than the red elderberry. The fruit of all elderberries is easy to pick quickly because it grows in clusters.

Miscellaneous Fruits

Several additional species of plants in the Northwest produce fruits that vary widely in their edibility and taste.

Crowberry

Lily Family (Liliaceae)

One-flowered Clintonia *Clintonia uniflora*

OTHER NAMES Queen's cup, blue bead, bead lily.
DESCRIPTION Perennial herb from rhizomes. **Height:** To 6" (15 cm) tall. **Leaves:** Basal, lance-shaped; to 8" (20 cm) long. **Flowers:** White, single; to 1" (2.5 cm) diameter (May to August). **Fruit:** Blue, berry-like; to ³⁄₈" (9 mm) diameter (July and August).
HABITAT Moist forests, open sites.

RANGE Alaska to California, east to Alberta, Colorado and Utah.
EDIBILITY Considered inedible by some, and edible but tasteless by others, but not considered poisonous.
NOTES To view the delicate beauty of the white blossoms or the blue, bead-like fruit of this species is always a pleasant experience while enjoying a walk in the woods. The fruit of queen's cup provides food for ruffed grouse and probably other birds. The leaves are also food for elk. Native people used this plant to make a blue dye and to treat sore eyes and dog bites.

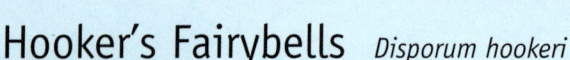

Hooker's Fairybells *Disporum hookeri*

OTHER NAMES Oregon fairybells; formerly *Prosartes hookeri*.
DESCRIPTION Perennial herb with hairy stems from rhizomes. **Height:** To 40" (1 m) tall. **Leaves:** Alternate, veins parallel, hairs on margins point forward; to 5" (12.5 cm) long. **Flowers:** Greenish white and bell-shaped, style has hairy surface; to ³⁄₄" (1.9 cm) long (May and June). **Fruit:** Green ripening to yellow or red, smooth or hairy, egg-shaped berry; to ½" (1.2 cm) across (August).
HABITAT Moist woods.
RANGE British Columbia to California, east to Idaho and Montana.
EDIBILITY Edible, juicy and somewhat sweet.
NOTES The bright red fruit of Hooker's fairybells was eaten by some Native people; others considered it to be poisonous. The fruit is not abundant, however, so it is not often collected for food.

SIMILAR SPECIES
Rough-fruited Fairybells *Disporum trachycarpum*
The flowers of rough-fruited fairybells have stamens that hang below the petals and a smooth style. Its red fruit has a velvety, somewhat warty surface. The fruit is edible and tastes distinctly like apricot. This species grows in moist woods from British Columbia to California, east to Alberta and to New Mexico.

Lily Family (Liliaceae)

False Solomon's-seal *Maianthemum racemosum*

OTHER NAMES Solomon-plume, false spikenard; formerly *Smilacina racemosa*.
DESCRIPTION Perennial herb from rhizomes. **Height:** To 40" (1 m) tall. **Leaves:** Alternate, elliptical, parallel veins; to 8" (20 cm) long. **Flowers:** Creamy white, to 1/8" (3 mm) long, small terminal pyramidal cluster (May and June). **Fruit:** Reddish, round berries, often spotted with purple; to 9/32" (7 mm) diameter (August to October).
HABITAT Moist woods, stream banks; low to subalpine elevations.
RANGE Alaska to southern California.
EDIBILITY The fleshy berries are edible but not especially tasty.
NOTES This delightful species is very showy. Its flowers grow in a "fluffy" raceme and smell strongly of perfume, and they are eventually succeeded by spotted fruit. Native people used various parts of this plant to treat several ailments, including rheumatism and kidney troubles. The extremely sweet fruit, known in some areas as sugarberry, tastes somewhat like saccharin.

Star-flowered False Solomon's-seal
Maianthemum stellatum

OTHER NAMES Star-flowered Solomon's seal, false Solomon's-seal; *M. stellata, Smilacina stellata*.
DESCRIPTION Perennial herb from rhizomes. **Height:** To 2' (60 cm) tall. **Leaves:** Alternate, lance-shaped; to 6" (15 cm) long. **Flowers:** Cream-colored, star-like terminal clusters of 5–10; to 1/4" (6 mm) long (May to early June). **Fruit:** Greenish yellow striped with purple, maturing to dark blue or black, round; to 3/8" (9 mm) diameter (July to September).
HABITAT Moist woods, clearings; low to subalpine elevations.
RANGE Alaska to Colorado, east to Arizona and Nevada.
EDIBILITY The fruit is edible, sweet but not especially tasty.
NOTES The Ktunaxa people of British Columbia considered both the star-flowered false Solomon's-seal and false Solomon's-seal (see above) to be the favorite food of the grizzly bear, but inedible for humans.

Lily Family (Liliaceae)

Wild Lily-of-the-valley *Maianthemum canadense*

OTHER NAMES Canada mayflower, bead ruby, Canada beadruby, two-leaved Solomon's Seal.
DESCRIPTION Perennial herb from rhizomes. **Height:** To 10" (25 cm) tall. **Leaves:** Alternate, heart-shaped to oval; to 3" (7.5 cm) long. **Flowers:** White, 4 segments (rather than the usual 6), clusters; to ¼" (6 mm) across (April to June). **Fruit:** Green berry, ripening to red; to ³⁄₁₆" (5 mm) diameter (July and August).
HABITAT Moist woods and clearings.
RANGE Northwest Territories, northern British Columbia and Alberta, south to Montana and Wyoming, east to the Atlantic coast.

EDIBILITY Not edible. **Caution:** The berries are considered to be a strong purgative, and one authority suggests that they may contain glycosides (heart stimulants).
NOTES Wild lily-of-the-valley is common throughout eastern North America, but in the West it occurs only in northern ranges. It is a delicate species that sometimes forms a thick plant cover on the moist forest floor, especially near aspen and pine. Its red fruit is occasionally eaten by grouse, deer mice and chipmunks. It is called wild lily-of-the-valley because it strongly resembles a European species, the true lily-of-the-valley (*Convallaria majalis*).

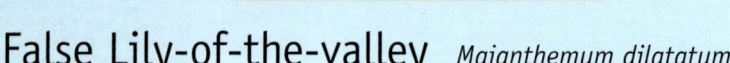

False Lily-of-the-valley *Maianthemum dilatatum*

OTHER NAMES Deerberry, wild lily of the valley.
DESCRIPTION Perennial from a creeping rootstock. **Height:** To 16" (40 cm) tall. **Leaves:** Alternate, heart-shaped; to 4" (10 cm) long. **Flowers:** Creamy white, in elongated terminal cluster, flower parts in 4's (normally 3's in lily family); to ⅛" (3 mm) long (June). **Fruit:** Green berry, mottled with brown, maturing to red, to ¼" (6 mm) diameter (July and August).
HABITAT Moist forests, clearings and meadows.
RANGE Alaska to California, east to Idaho.
EDIBILITY Edible but not tasty. Native people are known to have eaten the fruit, but it was not highly regarded.

NOTES This plant favors shady locations, where it often spreads into a thick, lush covering over the forest floor, carpeting the ground with its shiny leaves. Aboriginal people traditionally used this plant in a wide variety of ways, including treatment of cuts, sore eyes, internal injuries and sterility.

Lily Family (Liliaceae)

Clasping Twistedstalk *Streptopus amplexifolius*

DESCRIPTION Perennial herb from rhizomes. **Height:** To 40" (1 m) tall. **Leaves:** Alternate and elliptical, clasping stem at base; to 6" (15 cm) long. **Flowers:** Greenish white, elongated bell shape on kinked stalk; to ½" (1.2 cm) long (May to August). **Fruit:** Yellow to red or dark purple oval berries; to ⅜" (9 mm) long (August and September).
HABITAT Rich forests, open areas, stream banks and similar areas; low to subalpine elevations.
RANGE Alaska to California and across much of eastern North America.

EDIBILITY The berries are edible but not highly rated, and can cause diarrhea in humans. Grouse and chipmunks find them much more agreeable.
NOTES The name twistedstalk originates from the distinctive zigzag pattern along the branched stems. Most Native people regarded the entire plant, including the fruit, to be poisonous, but some groups ate the fruit and young plants raw as well as cooked in stews and soups. Aboriginal people referred to this plant by a variety of local names, including frog berries, black bear berries, witch berries and owl berries.

SIMILAR SPECIES
Rosy Twistedstalk
Streptopus roseus
This plant grows to a height of 2' (60 cm). Its distinctive, colorful rosy flowers hang from stalks that are not kinked. The cherry red fruit is said by some to have laxative properties. This species grows from Alaska to Oregon, east across Canada and to Minnesota and Georgia.

Small Twistedstalk *Streptopus streptopoides*
This species grows to 8" (20 cm) tall, with flowers that are rose to greenish purple, and red fruit. It is found in coniferous forests at mid- to subalpine elevations from Alaska to Washington.

Sandalwood Family (Santalaceae) & Rose Family (Rosaceae)

Northern Comandra *Geocaulon lividum*

OTHER NAMES Bastard toad-flax, timberberry, pumpkin berry, dogberry; also known as *Comandra livida*.
DESCRIPTION Perennial plant from a creeping rootstock. **Height:** To 8" (20 cm) tall. **Leaves:** Alternate, oval to oblong; to 1 1/8" (3 cm) long. **Flowers:** Green, close to main stem, in clusters of 1–4; to 5/8" (1.5 cm) long (June and July). **Fruit:** Orange, fleshy, round; to 5/16" (8 mm) diameter (July and August).
HABITAT Dry open forests, bog forests; low to medium elevations.
RANGE Alaska to northern Washington.
EDIBILITY Edible but tasteless fruit.
NOTES This parasitic plant obtains nutrients from the roots of other plants to supplement the food it produces. Although its attractive fruit is edible, Native peoples rarely ate it. They used various parts of the plant as remedies for chest problems and other ailments, and as a poultice for wounds.

Indian-plum *Oemlaria cerasiformis*

OTHER NAMES Indian plum, oso-berry, bird cherry, skunk bush; formerly *Osmaronia cerasiformis*.
DESCRIPTION Deciduous shrub. **Height:** To 17' (5 m) tall. **Leaves:** Alternate; to 4 3/4" (12 cm) long. **Flowers:** Greenish white, hanging clusters; flowers emerge before leaves; to 3/8" (9 mm) long (March and April). **Fruit:** Yellow ripening to dark blue with a whitish bloom (powdery covering), plum-shaped; to 3/8" (9 mm) across, large stone present (June).
HABITAT Open woods, wet and dry areas; low elevations.
RANGE Southern British Columbia to central California.
EDIBILITY Edible but "puckery," with a large stone and only a thin layer of flesh. This fruit tastes best just as it is turning from red to purple.

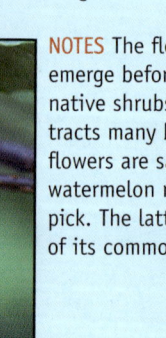

NOTES The flowers of Indian-plum emerge before the flowers of all other native shrubs. The fruit of this plant attracts many birds, which feed on it. The flowers are said to smell somewhat like watermelon rind or cat urine—take your pick. The latter may be the origin of one of its common names: skunk bush.

Crowberry family (Empetrraceae) and Ginseng Family (Araliaceae)

Crowberry *Empetrum nigrum*

OTHER NAMES Mossberry, curlew berry, crake berry, black berry.
DESCRIPTION Creeping, low-lying shrub. **Height:** To 8" (20 cm) tall. **Leaves:** Green to wine-colored, evergreen, needle-like; to 1/4" (7 mm) long. **Flowers:** Crimson, inconspicuous, in leaf axils; to 1/4" (6 mm) long (May to early June). **Fruit:** Black, shiny, berry-like drupes; to 3/8" (9 mm) across (August and September).
HABITAT Bogs, low-lying coastal headlands, moist shady forests; sea level to alpine.
RANGE Alaska to Newfoundland and along the Pacific coast to northern California.
EDIBILITY Edible. Can be eaten raw or cooked. Large white inedible seeds are also present. This fruit has a very low pectin content, but it is often used to make jelly and pies in the North, where it is plentiful.
NOTES Crowberry was an important fruit to northern aboriginal peoples, including the Inuit. It is rated by some as one of the most important fruits to northerners, second only to cloudberry (see p. 36). The taste of its berries varies greatly with its habitat, so it gets mixed reviews. Some say the berries taste like turpentine; others rave about their sweet taste; still others describe it as an acquired taste. The fruit tastes better if it is picked after the first frost, or if sugar and lemon are added.

Wild Sarsaparilla *Aralia nudicaulis*

DESCRIPTION Woody perennial from a rhizome base. **Height:** To 2' (60 cm) tall. **Leaves:** Single and compound, with 3 major divisions each of which may have 3–5 leaflets, toothed edge; to 20" (50 cm) long. **Flowers:** Greenish white umbels or ball-shaped clusters, each flower with 5 petals; to 1/8" (3 mm) long (May). **Fruit:** Greenish white changing to dark purple or black when ripe, forms a drupe or ball-shaped cluster; to 1/4" (6 mm) diameter (July).
HABITAT Shaded forests, especially mixed-wood stands; low elevations.
RANGE Across Canada to northeastern Washington, Montana, Colorado, Michigan and the eastern U.S.
EDIBILITY Edible. A large seed is present and the berries have little flavor. The fruit has also been used to make beer and wine.

NOTES Wild sarsaparilla often forms an impressive understory layer in a variety of forest types. The fruit is produced sporadically and was used as a flavoring by early North American settlers. This species got its name from a similar-tasting but unrelated tropical species, prickly vine sarsaparilla (*Smilax officinalis*). Native people in the Bella Coola, B.C., area made a drink by boiling the rhizomes of wild sarsaparilla in water. The plant has also been used medicinally for a range of ailments.

Ginseng Family (Araliaceae)

Devil's Club *Oplopanax horridus*

OTHER NAME Formerly *Echinopanax horridum*.
DESCRIPTION Deciduous shrub armed with numerous large spines to 3/8" (9 mm) long. **Height:** To 10' (3 m) tall. **Leaves:** Alternate, maple leaf-shaped with 5–7 lobes; to 14" (35 cm). **Flowers:** Greenish to white, many on an upright raceme; raceme to 3/4" (2 cm) long (May to July). **Fruit:** Red, flattened, showy clusters in a pyramidal raceme; to 5/16" (8 mm) long (July to September).
HABITAT Moist, open areas, forest edges and stream banks.
RANGE Alaska to Yukon Territory, south to Oregon and east to Idaho and Montana.
EDIBILITY Inedible for humans. Bears, however, relish the fruit.

NOTES Devil's club is named for the formidable spines that cover its branches and the undersides of its large leaves. Scratches from these spines can cause allergic reactions, and spines that break off and get embedded in the skin can cause infection.

Devil's club was an important plant species to many Native peoples, who used it in the treatment of ailments such as rheumatism, arthritis, ulcers, diabetes and digestive troubles. Its roots and inner bark have therapeutic properties. The plant is still used today in the treatment of diabetes.

Oleaster Family (Elaeagnaceae) and Buckthorn Family (Rhamnaceae)

Silverberry *Elaeagnus commutata*

OTHER NAME Wolf-willow.
DESCRIPTION Deciduous shrub. **Height:** To 10' (3 m) tall. **Leaves:** Opposite, silvery, oval; to 4" (10 cm) long. **Flowers:** Yellow, funnel-shaped; to ⅝" (1.6 cm) long (June and July). **Fruit:** Silver, oval; contains a large nutlet; to ⅜" (9 mm) long (August to October).
HABITAT Well-drained open sites, stream banks and similar areas.
RANGE Alaska to Quebec, south to Idaho, South Dakota and Minnesota.

EDIBILITY Edible but far from incredible. Some say the fruit of silverberry makes a very nice jam.
NOTES This plant's silver leaves and berries are distinctive, as are the musky-smelling flowers. The large nutlets in the fruit were sometimes used as decorative beads. Silverberry was considered to be a famine food by some Native people, but other northern tribes ate it regularly, mixed with various foods.

Cascara *Rhamnus purshiana*

DESCRIPTION Deciduous shrub or small tree. **Height:** To 33' (10 m) tall. **Leaves:** Alternate, oblong to egg-shaped; to 5" (12 cm) long. **Flowers:** Greenish yellow, umbrella-shaped clusters of 8 to 50; to 3/16" (4 mm) long (May and June). **Fruit:** Black, berries; to 5/16" (8 mm) diameter (August and September).
HABITAT Wet to dry sites, shady areas, often in mixed woods; low to medium elevations.
RANGE Coastal regions, southern British Columbia to California.
EDIBILITY Edible but seedy. **Caution:** This fruit can be a severe purgative.

NOTES The bark of cascara is well known as the source of cascara sagrada, a drug that has been used as a laxative for more than a hundred years. This shrub was so extensively harvested in the past—up to 1,500 tons (1,350 tonnes) of bark annually—that it is not often found outside parks and similar protected areas today. It is hoped that over time the various game birds and songbirds that feed on the berries will help to propagate this species.

Honeysuckle Family (Caprifoliaceae)

Red Twinberry *Lonicera utahensis*

OTHER NAME Utah honeysuckle.
DESCRIPTION Deciduous shrub. **Height:** To 6½′ (2 m) tall. **Leaves:** Opposite, elliptical to oblong; to 3″ (7.5 cm) long. **Flowers:** Light yellow, tubular, in pairs; to 1″ (2.5 cm) long (May to July). **Fruit:** Bright red, paired, one flower often smaller than the other, round; to ½″ (1 cm) diameter (July and August).
HABITAT Moist to wet areas, open forests; medium to subalpine elevations.
RANGE British Columbia to California, east to Alberta, Montana, Wyoming and Utah.
EDIBILITY Edible and juicy. This species is not often found in abundance, so it makes a refreshing treat during a trail hike.
NOTES The flowers of this species, like those of other honeysuckles, are favorites for hummingbirds. The twin flowers are replaced in the summer by distinctive twin red berries, which make the plant easy to identify. This honeysuckle has been used as a mild laxative and as a wash for sores and infections. It grows east of the Cascades.

Honeysuckle Family (Caprifoliaceae)

Black Twinberry *Lonicera involucrata*

OTHER NAME Bracked honeysuckle.
DESCRIPTION Deciduous shrub. **Height:** To 10′ (3 m) tall. **Leaves:** Opposite, elliptic to lance-shaped, tips pointed; to 4″ (10 cm) long. **Flowers:** Yellow, trumpet-shaped, in pairs with green to purplish bracts; to 3/4″ (2 cm) long. (May to July) **Fruit:** Shiny black berries, in pairs; to 15/16″ (8 mm) across (July to September).
HABITAT Moist woodlands and thickets; sea level to high elevations.
RANGE Alaska to California, east to Alberta, Montana and New Mexico.
EDIBILITY Bitter and inedible.

NOTES Native people harvested this honeysuckle for a variety of medicinal uses. It has been used to treat sore throats, broken bones and sores, as well as arthritis.
The fruit of this shrub is not a source of food for man but is said to be eaten by bears. The cedar waxwing, Swainson's thrush and other birds also feed on the berries. In the springtime, hummingbirds sip nectar from the yellow flowers of this shrub.

SIMILAR SPECIES

Western Trumpet Honeysuckle *Lonicera ciliosa*
This climbing vine grows to a height of 30′ (9 m). Its terminal leaves form a disk that contains trumpet-shaped orange flowers. Later in the season, red to orange fruit is produced in terminal clusters. It is not edible. Western trumpet honeysuckle grows from British Columbia to California, east to Montana.

Oleaster Family (Elaeagnaceae) & Heath Family (Ericaceae)

Canada Buffaloberry *Shepherdia canadensis*

OTHER NAMES Buffaloberry, soapberry, soopolallie.
DESCRIPTION Deciduous shrub. **Height:** To 20′ (6 m) tall. **Leaves:** Opposite, oval; to 2 3/8″ (6 cm) long. **Flowers:** Yellowish brown, male and female flowers on separate plants; to 3/16″ (4 mm) diameter (April and June). **Fruit:** Bright red to yellow berries; to 1/4″ (6 mm) long (July and August).
HABITAT Forest edges and stream banks.
RANGE Alaska to Newfoundland, south to New Mexico.
EDIBILITY Edible but bitter. Can be eaten raw, or cooked as a jelly or drink. **Caution:** The fruit of this shrub contain saponin, which can irritate the stomach. Eating large amounts may cause diarrhea, vomiting and cramps.

NOTES Native people collected the berries of Canada buffaloberry easily by beating the branches under which a hide or canvas was spread. This fruit was often made into "Indian ice cream" by whipping the berries and a bit of water with strands of grass or bark. More recently, sugar was added to this frothy, meringue-like mixture. Like egg whites, these berries will only whip in bowls that are made of glass, porcelain or metal—plastic and rubber dishes contain grease or oil residues, which inhibit frothing. The berries were also traditionally added to stews, cooked into a sauce for buffalo steaks and, as one of its common names suggests, used as soap.

Arbutus *Arbutus menziesii*

OTHER NAMES Madrone, Pacific madrone.
DESCRIPTION Deciduous tree with peeling bark. **Height:** To 100′ (30 m) tall. **Leaves:** Evergreen, alternate; to 6″ (15 cm) long. **Flowers:** White, urn-shaped, clusters; to 1/4″ (7 mm) long (March to May). **Fruit:** Orange-red berries, fine granular surface; to 3/8″ (9 mm) diameter (September to November).
HABITAT Dry slopes, rocky areas, often associated with Douglas-fir and Garry oak; low to mid-elevations.
RANGE Coastal regions, southern British Columbia to California.
EDIBILITY The fruit is edible raw, boiled or steamed, although it is mealy, bland and occasionally bitter. Eating more than a few berries will cause stomach cramps in humans. However, the fruit of the arbutus is popular with the band-tailed pigeon, varied thrush and other birds.

NOTES The arbutus is well known for its smooth, peeling bark, which changes colors with the seasons. In spring it is bright green, then it transforms to red and eventually peels off entirely. This tree is not well known for its fruit and was not often eaten by humans, except in California. The bark and leaves were used for their medicinal properties.

Heath Family (Ericaceae)

Common Bearberry *Arctostaphylos uva-ursi*

OTHER NAME Kinnikinnick.
DESCRIPTION Trailing evergreen shrub. **Height:** To 6" (15 cm) tall. **Leaves:** Alternate, oval; to 1 1/8" (3 cm) long. **Flowers:** White to pink, urn-shaped, drooping clusters; to 3/16" (4 mm) long (May and June). **Fruit:** Red, berry-like drupes; to 1/2" (1 cm) across (August and September).
HABITAT Well-drained, open woodland sites and clearings; low to alpine elevations.
RANGE Circumpolar. Alaska to Labrador, south to California and New Mexico.
EDIBILITY Edible, but far from incredible.
NOTES The fruit of common bearberry is edible and often lingers on the branches through much of the winter. It is a mealy, somewhat tasteless fruit, but was an important survival food for many Native people. Sugar was often added to improve the taste. The fruit can be used in making jams, jellies, pies and wines.

In the last two centuries, common bearberry was mixed with the bark of red-osier dogwood (p. 44) and smoked by some Native people. The Nuxalk used the stems of the gooseberry to make pipes in which dried bearberry leaves were smoked. A wide range of birds and mammals enjoy eating the fruit of the bearberry, including bears, as the name suggests.

Several species of *Arctostaphylos* are found throughout the region. The most common of these are listed below.

SIMILAR SPECIES

Black Alpine Bearberry *Arctostaphylos alpina*
This dwarf shrub, sometimes referred to as whortleberry, has wrinkled leaves that turn bright red in autumn. Its solitary, yellow to yellowish green flowers produce black fruit. It is a circumpolar species that grows in the alpine tundra from Alaska across the Northwest Territories and south to northern British Columbia. The mealy fruit is edible and makes an excellent survival food. Another similar species, red alpine bearberry (*Arctostaphylos rubra*), whose fruit is red, may also be encountered. It ranges from the Rockies east to the Atlantic coast.

Hairy Manzanita *Arctostaphylos columbiana*
This species, an evergreen shrub whose young leaves and twigs are hairy, produces mealy, blackish red fruit. It grows to a height of 10' (3 m). Hairy manzanita is found in dry openings along the coast from southern British Columbia to California.

Heath Family (Ericaceae)

Salal *Gaultheria shallon*

OTHER NAME Laughing berries.
DESCRIPTION Evergreen deciduous shrub that may creep low to the ground or grow erect. **Height:** To 16′ (5 m) tall. **Leaves:** Evergreen, alternate, leathery, with a shiny surface. To 4″ (10 cm) long. **Flowers:** White to pink, urn-shaped, in elongated clusters of 5–15 at the branch ends; to 3/8″ (9 mm) long (May to early June). **Fruit:** Blue to black, "berries" are actually fleshy sepals, in elongated clusters of 5–15 at the branch ends; to 3/8″ (9 mm) across (July and August).
HABITAT Various coniferous forests; low to medium elevations.
RANGE Coastal regions, southeast Alaska to southern California.
EDIBILITY Edible and choice, although dry areas sometimes produce tasteless fruit. It can be eaten raw or made into jams, jellies, syrups, pies, pancakes and other foods.
NOTES Salal is a shrub that grows in abundance in many areas along the coast. Its berries are juicy with good flavor. This was the most important fruit for many Native peoples on the coast from Alaska to California. It was eaten raw or dried into cakes, and often traded or sold with other fruit. The cakes, which weighed as much as 15 lbs (6.8 kg), were dipped in seal or whale oil as they were eaten. Some groups, such as the Haida of British Columbia, ate the dried berries with the oil of the eulachon, a small fish that is abundant at certain times of the year. The berries contain natural sugars and mix well with other foods. Some people have described this fruit as an almond-flavored blueberry. At Fort Victoria, B.C., in colonial times, Hudson's Bay Company employees made wine from salal berries. David Douglas, a naturalist based at Fort Vancouver during the 1820s, was so impressed by salal that he collected seeds and took them home to Britain in 1828. Today salal is available for gardens from nurseries, and sprays of salal leaves are added to many bouquets of cut flowers.

Poisonous Fruits

A few wild fruits that grow in the West are poisonous. It is important to identify them, because just a few berries of some species can cause serious illness or even death. Remember too that fruit is not necessarily safe for humans just because it is eaten by wildlife—the metabolism of wild birds and animals is much different than ours. Harvest and eat only those species you know are safe after consulting an expert.

Baneberry

Yew Family (Taxaceae) & Honeysuckle Family (Caprifoliaceae)

Western Yew *Taxus brevifolia*

OTHER NAMES Pacific yew, mountain mahogany.
DESCRIPTION Evergreen tree. **Height:** To 80' (24 m) tall. **Leaves:** Flat needles; to 1½" (3.5 cm) long. **Flowers:** Male: cylindrical cones with yellowish pollen sacs. Female: Minute green flowers to ⅛" (3 mm) long (June). **Fruit:** Salmon to scarlet, cup-shaped, covers a large brown seed; to ⅜" (9 mm) across (August to October).
HABITAT Moist mature forest, common along the coast; low to medium elevations.
RANGE Southern Alaska to Idaho, south to California.
EDIBILITY **Caution:** All parts of this plant, including the fruit, are considered **toxic**. The poison is an alkaloid that alters the rhythm of the heartbeat and can cause sudden heart failure.

NOTES The bark of this shade-loving tree contains taxol, a substance that is currently being tested to treat a range of cancers, including ovarian, breast and kidney cancer. Birds are attracted to the fruit of this species, and this is one of the main ways in which the tree distributes its seeds.
Native people used this plant as a medicine in several ways. The Haida of B.C. made a sedative from the needles and twigs, and the Mendocino of California made a poison. The wood of the western yew, which is extremely hard and durable, was used to make many useful items, such as dishes, spoons, boxes, clubs, paddles, fire tongs and snowshoe frames.

Common Snowberry *Symphoricarpos albus*

OTHER NAME Waxberry.
DESCRIPTION Deciduous shrub from rhizomes, twigs hollow and hairless. **Height:** To 9' (2.75 m) tall. **Leaves:** Opposite, oval; to ¾" (2 cm) long. **Flowers:** Pinkish, bell-shaped; to ¼" (7 mm) long (June). **Fruit:** White, round, berry-like, in clusters; to ⅝" (1.5 cm) diameter (August).
HABITAT Moist clearings, stream banks, thickets; low to medium elevations.
RANGE Common throughout most of North America.
EDIBILITY **Poisonous. Caution:** All parts of this plant, including the fruit, are considered **toxic**.
NOTES A total of 10 species of snowberries grow in North America, and another occurs in China. They are attractive, colorful plants in our gardens during fall and winter. They also provide food for grouse, thrushes, grosbeaks and other birds, as well as bear, beaver, hare, deer and other mammals.

SIMILAR SPECIES
Creeping Snowberry *Gaultheria hispidula*
This small, creeping evergreen shrub grows to 5½" (14 cm) long with the characteristic white fruit. It can be found in bog habitats from British Columbia to Labrador, south to Washington and east to Idaho. Like the common snowberry, this plant is **toxic**—do not eat the fruit or any other part of it.

Buttercup Family (Ranunculaceae)

Baneberry *Actaea rubra*

OTHER NAMES Snake berry, doll's eyes, red and white baneberry; includes *A. arguta*, *A. eburnea*.
DESCRIPTION Perennial from rhizomes. **Height:** To 39" (1 m) tall. **Leaves:** Alternate, 2 or 3 times pinnate into threes; to 4" (10 cm) long. **Flowers:** White, rounded clusters; to 3/16" (4 mm) long (April to June). **Fruit:** Glossy red or white, round; to 3/8" (8 mm) diameter (August and September).
HABITAT Moist shady woods, stream banks and similar areas; low elevations to subalpine.
RANGE Alaska to California, east to Idaho.
EDIBILITY Poisonous. Caution: Severe cramps, headache, vomiting or dizziness can be caused by eating as few as 2 baneberries.
NOTES All parts of this plant are poisonous, especially the berries and roots—they are known to suppress the vagus nerve, which can cause cardiac arrest. Native peoples have carefully used this species in very small amounts to treat uterine pains, stomach problems and other ailments. It was also used to treat sick horses. The fruit is commonly seen in both white and red color phases.

Holly Family (Aquifoliaceae)

English Holly *Ilex aquifolium*

DESCRIPTION Evergreen tree.
Height: To 50′ (15 m) tall.
Leaves: Glossy, alternate, leathery, spiny teeth; to 2½″ (6 cm) long. **Flowers:** White or green, bell-shaped, clusters; to ¼″ (6 mm) long (May). **Fruit:** Red, round; to ⅜″ (9 mm) diameter (September to December).
HABITAT Moist forests and edges.
RANGE Naturalized from British Columbia to California, west of the Cascades.
EDIBILITY Poisonous. Caution: These berries can trigger violent vomiting.
NOTES The leaves of the English holly, a species introduced from Europe, resemble those of our native Oregon grape (see p. 11). The fruit of this species, which is poisonous to humans, stays on the trees throughout the winter, providing food for various birds. The fragrant flowers are followed by fruit, but only on female trees. Purchase 2 plants, a male and a female, to produce the colorful fruit in the garden—but only if there is no danger that children will eat the berries.

Sumac Family (Anacardiaceae)

Poison-ivy *Toxicodendron radicans*

OTHER NAME Formerly *Rhus radicans*.
DESCRIPTION Deciduous shrub from creeping rhizomes, can also be a clambering plant.
Height: To 6½′ (2 m) tall. **Leaves:** Compound with 3 pointed leaflets; to 4″ (10 cm) wide.
Flowers: Whitish yellow, round, clusters; to ⅛″ (3 mm) diameter (May and June). **Fruit:** Green turning white, berry-like, round; to ¼″ (6 mm) diameter (July and August).
HABITAT Dry and moist sites, often on the edges of trails; low elevations.
RANGE British Columbia to Nova Scotia, south to Oregon; more common in eastern North America.
EDIBILITY Poisonous. Caution: All parts of this plant, including the fruit, are **toxic**. Caution is advised.
NOTES Contact with any part of poison-ivy causes an itching, burning rash that persists for many

days. If you do brush up against this plant, wash with several changes of water. If itching persists, apply camomile lotion to the area affected up to 12 hours after contact, which may prevent the rash. If a severe rash develops, contact a physician. All parts of the poison-ivy plant contain the volatile oil urushiol, highly irritating to most people. Urushiol is a very stable compound-some individuals have had reactions to dried poison-ivy specimens over a century old!
Everyone should avoid this species, because even those who are not allergic to it can transfer the oil to others who are allergic. Its toxic properties are so strong that if the plant is burned, the pollen and smoke will cause allergic reactions and can even be fatal for highly allergic people.

SIMILAR SPECIES

Poison Oak *Toxicodendron diversilobum*
This species, formerly *Rhus diversiloba*, also has a nasty reputation. As its common name suggests, the plant displays oak-shaped leaves. Its flowers emerge in drooping clusters that form white berry-like drupes. All parts of this plant are poisonous and should be avoided. Poison oak can be found along the Pacific coast from southern British Columbia to Oregon.

Nightshade Family (Solanaceae)

European Bittersweet *Solanum dulcamara*

OTHER NAMES Bittersweet, woody nightshade.

DESCRIPTION Semi-woody, climbing vine from a rhizome. **Height:** To 10′ (3 m) tall. **Leaves:** Alternate; to ⅜″ (8 mm) long. **Flowers:** Purple, shooting star-like; to ⅝″ (1.6 cm) across (April to October). **Fruit:** Shiny green changing to yellow and bright red, round to oval; to ⅜″ (9 mm) diameter (April to October).

HABITAT Roadsides, waste areas, moist clearings and forest edges; low elevations.

RANGE British Columbia to California, east to Utah.

EDIBILITY Caution: All parts of this species are **poisonous**. The plant produces a toxic alkaloid, solanine, in its unripe fruit and leaves. Solanine disrupts the gastrointestinal tract and nervous system. Sickness and **death** have been reported in both animals and humans who consumed this plant.

NOTES This species, introduced from Europe, has been naturalized across much of temperate North America. It scientific name stems from two words: *dulca*, which means "sweet," and *mara*, which means "bitter." The flowers and fruit appear together. Various birds feed on the European bittersweet, including the song sparrow.

Freezing, Canning and Drying

Fresh fruit has a short shelf life, so it must be preserved. Today it is much easier to store fruit than it used to be. In the past, Native people dried fruit into cakes or submerged them in a mixture of water and oil. They also mixed fruit with other foods, including fat and meat, to make pemmican. The three ways to store fruit are freezing, canning and drying. Each of these has its advantages and disadvantages.

Freezing Wild Fruit

Freezing is the easiest way to store fruit. Choose prime ripe fruit for freezing, and wash and drain it either before or after freezing. (Some fruit, such as raspberries, may be too soft for washing while fresh and should be washed after freezing.) Fruit is best quick-frozen by spreading it out on trays in the freezer for about 2 hours before placing it in airtight freezer bags or containers. Unwashed fruits will not stick together when frozen. Frozen fruit keeps well for as long as 12 months. Follow the freezer manufacturer's instructions.

Canning Wild Fruit

Canning is a good way to store fruit that can be collected in volume, such as blueberries and saskatoons. Choose fruit that is fresh, firm and field-ripened, then wash and drain it before preparing it. Specific methods for canning fruit vary depending on the type of fruit. For detailed instructions, consult a book or Internet site with home-canning tips and techniques.

Drying Wild Fruit

Many wild fruits can be preserved easily by drying. Firm fruits such as blueberries can be dried separately, like raisins, but juicier species are best mashed, poured out and dried as fruit leather (see p. 87). Seedy fruits, such as raspberries, do not dry as well as others, but some of them make good fruit leather. Elderberries should be cooked before they are dried, because the raw berries are toxic to many people.

Fruit with a hard outer skin should be blanched for 1 minute before being dried. This adds moisture to the skin so that the fruit dries evenly inside and out. Lay the fruit on paper towels after blanching and before drying. Fruit can be dried slowly in an oven set at very low heat—140°F (60°C) or lower. Leave the door partly open to allow moisture to escape. This process can take as long as 8 hours. If fruit is too abundant and ripe, it may be pureed and frozen until you have time to dry it as fruit leather (see p. 87).

Fruit may also be dried in the sun in extremely sunny, arid areas. Spread it out carefully in the sun during the day and bring it in at night. It can take as long as 2 days to dry fruit this way.

Food dehydrators come in many types and sizes. Follow the manufacturer's instructions, and make sure all of the fruit is about the same size so that it all dries at the same rate. Dried fruit can be stored for 12 months or longer, although it loses color, flavor and nutritive value over time. When you are ready to use the fruit, soak it in water overnight and use it as you would fresh fruit.

Recipes
Tried and True

This section was prepared by Susan Servos-Sept

BLACKBERRIES

Blackberry Loaf

A simple-to-make dessert loaf.

2½ cups (625 ml) all-purpose flour
1 cup (250 ml) sugar
3½ teaspoons (17 ml) baking powder
1 teaspoon (5 ml) salt
3 tablespoons (45 ml) salad oil
¾ cup (175 ml) milk
1 egg
1 cup (250 ml) fresh blackberries

Preheat oven to 350°F (180°C). Grease and flour a 9x5" (23x12 cm) loaf pan. Place all ingredients in a large bowl and beat for 30 seconds on medium speed. Pour into loaf pan and bake 55 to 65 minutes. Remove loaf from pan while still slightly warm. Slice when cooled.

Sue's Blackberry Pie

Fabulous!

¼ cup (60 ml) water
1 cup (250 ml) sugar
2½ tablespoons (37 ml) cornstarch
5 cups (1.25 l) blackberries (fresh or frozen)
2 teaspoons (10 ml) lemon juice

9" (23 cm) unbaked pie shell
Lattice crust (optional)

Preheat oven to 350°F (180°C).
In a saucepan, combine water, sugar and cornstarch. Mix well. Add berries and lemon juice, and bring to a boil. Continue to boil gently, stirring constantly, until filling has thickened. Remove from heat and pour filling into pie shell. Decorate top with lattice crust if desired. Bake for 50 to 60 minutes.

RASPBERRIES

Raspberry Cheese Pie

A delicious dessert.

Crust
10" (25 cm) uncooked pie shell

First Layer (bottom filling)
8 oz (250 g) cream cheese, softened
1 egg
½ cup (125 ml) icing sugar
11 oz (300 ml) sweetened condensed milk
¼ cup (60 ml) lemon juice

Second Layer (top filling)
½ cup (125 ml) sugar
2 tablespoons (30 ml) cornstarch
¼ cup (60 ml) water
2 cups (500 ml) fresh (or frozen and thawed) raspberries
1½ teaspoons (3 ml) butter
¼ teaspoon (1 ml) almond extract

Topping
⅔ cup (150 ml) all-purpose flour
⅓ cup (75 ml) brown sugar
¼ teaspoon (1 ml) cinnamon
¼ teaspoon (1 ml) salt
¼ cup (60 ml) butter
½ cup (125 ml) chopped pecans

Preheat oven to 375°F (190°C).

First Layer (bottom filling): Beat cream cheese, egg and icing sugar in a bowl. Add condensed milk and lemon juice and beat well. Spread filling in pie shell.

Second Layer (top filling): In a saucepan, combine sugar, cornstarch and water. Mix well. Stir in raspberries. Cook on low heat until mixture thickens to the consistency of jam. Remove from heat and stir in butter and almond extract. Spread filling over bottom filling in pie shell.

Topping: In a separate bowl, mix flour, sugar, cinnamon and salt. Cut in butter until mixture is crumbly. Gently fold in chopped pecans. Sprinkle topping evenly over the top filling layer.

Bake for 45 to 55 minutes, or until layers are set and top is lightly browned.

Tart and Tangy Raspberry Lemonade

A refreshing summer drink.

¼ cup (60 ml) sugar
½ cup (125 ml) hot water
2½ cups (625 ml) cold water
½ cup (125 ml) lemon juice
¾ cup (175 ml) fresh raspberries
ice cubes

1 lemon slice for each glass
3 whole raspberries for each glass

Dissolve the sugar in the ½ cup (125 ml) of hot water. Let cool. Combine the cold water, lemon juice and raspberries in a blender. Pour into a glass, add ice and garnish with a lemon slice and a few raspberries.

Makes 4 - 8 oz (250 ml) glasses.

Variations
- For a sparkling drink, add ⅓ cup (75 ml) lemon-lime soda to each cup of lemonade.
- For a summer spritzer, add white wine.

BLUEBERRIES

Whole Wheat Blueberry Pancakes

A wholesome, fruity pancake.

¾ cup (175 ml) whole wheat flour
¾ cup (175 ml) all-purpose flour
3 teaspoons (15 ml) baking powder
½ teaspoon (2 ml) salt
1 tablespoon (15 ml) sugar
1 egg, well beaten
1¼ cup (265 ml) milk
2 tablespoons (30 ml) melted margarine
1 cup (250 ml) fresh blueberries

Combine dry ingredients in a large mixing bowl. In a separate bowl, combine the egg and milk. Add egg mixture to dry mixture and mix thoroughly. Stir in the melted margarine. Gently fold in blueberries. Spoon batter onto a hot greased frying pan. When bubbles appear on top of the cakes, flip and cook briefly on the other side.
Serve with syrup. Makes 8 - 6" (15 cm) pancakes.

NOTE: The sweetness of these pancakes will vary with the sugar content of the berries.

Blueberry Oatmeal Muffins

Moist, fruit-filled muffins.

1 egg
⅓ cup (75 ml) sugar
3 tablespoons (45 ml) butter or margarine, melted
1 cup (250 ml) milk
1 cup (250 ml) all-purpose flour
1 cup (250 ml) rolled oats
1 tablespoon (15 ml) baking powder
½ teaspoon (2 ml) salt
1 cup (250 ml) fresh blueberries
1 teaspoon (2 ml) cinnamon

Preheat oven to 400°F (200°C). Grease a muffin tin.
Combine egg, sugar and butter in a large mixing bowl. Add milk and beat well. In a separate bowl, combine flour, oats, baking powder cinnamon and salt. Mix well. Add flour mixture to to liquid mixture, stirring until just moistened. Gently fold in blueberries. Spoon batter into prepared muffin cups, filling each cup two-thirds full.
Bake 20 to 25 minutes or until muffins test done. Makes 12 muffins.

STRAWBERRIES

Wild Strawberry Smoothie

A wholesome, delicious drink.

1–1½ cups (250–375 ml) strawberries (fresh or frozen)
1 cup (250 ml) 2% milk
1 cup (250 ml) ice cubes (½ cup/125 ml if using frozen berries)
1 cup (250 ml) vanilla yogurt
1 tablespoon (15 ml) white sugar

Combine berries, milk and ice in a blender. Blend until frozen fruit and ice cubes are well chopped. Add yogurt and sugar. Blend just until mixture is thick, smooth and fluffy. Serve immediately. Makes 2 large servings.

Note: Any fruit may be substituted for strawberries.

Strawberry Cream Cheese Spread

A light, refreshing spread for crackers or bagels.

8 oz (250 g) cream cheese
¼ cup (60 ml) fresh ripe strawberries
2 teaspoons (10 ml) white sugar

Mash ingredients together until smooth and spreadable. Makes 1 cup (250 ml).

HUCKLEBERRIES

Red Huckleberry Glaze for Chicken

Coat skinless chicken breasts with this glaze and bake, broil or barbecue for a taste treat. The glaze can be also be used on pork and duck.

½ cup (125 ml) fresh red huckleberries
¼ cup (60 ml) brown sugar
1 tablespoon (15 ml) lemon juice
½ teaspoon (2 ml) grated orange zest

Place the huckleberries in a saucepan with just enough water to cover the bottom of the pan. Bring to a boil, then reduce heat and simmer uncovered until the berries are soft. Mash lightly, leaving some berries whole. Add sugar, lemon juice and orange zest. Bring to a boil once again. Simmer for 4 minutes and let cool. Pour over or brush onto chicken breasts and bake, broil or barbecue to taste.

SALAL

Salal Jelly

4 cups (1 l) salal berries
½ cup (125 ml) water
¼ cup (60 ml) lemon juice
1 - 57 g (2 oz) pkg powdered pectin
3 cups (750 ml) sugar

Crush berries and place in a saucepan with the water. Bring to a boil, stirring constantly. Strain mixture to remove skins and seeds. Pour juice into a heavy saucepan, add pectin and bring to a full boil, stirring constantly. Add sugar and lemon. Continue stirring and boil hard for 2 minutes. Remove from heat. Skim foam and pour jelly into hot sterile jars. Seal with canning lids. The jelly can also be placed in plastic containers for freezer storage.

MIXED BERRIES

Fruit Leather

A fruit treat to go.

4 cups (1 l) crushed fresh berries
¼ cup (60 ml) sugar

Combine berries of your choice with sugar in a saucepan. Bring to a boil and boil gently for 5 minutes, stirring constantly. Cool and put through a fine sieve to remove seeds and make the mixture a uniform consistency.

Pour mixture onto a non-stick baking sheet with edges, or one that is covered with parchment paper, to a thickness of ⅛–¼ inch (3–6 mm). Heat the oven to 140°F (60°C). Place fruit leather in the oven with the door slightly ajar to allow moisture to escape. Dry for 5 hours, or until the leather is dry to the touch and rolls easily off the baking sheet. Don't leave it for too long, or the leather will get tough.

Store in a paper bag for a week to allow further drying. Once dry, roll the fruit leather in wax paper or plastic wrap or store in an airtight container. It can also be frozen.

Fresh Fruit Parfait

One of the best ways to enjoy fresh wild fruit.

For each serving:
¾ cup (175 ml) fresh fruit of your choice
1 teaspoon (5 ml) sugar
cream to taste

Place fruit in a parfait glass. Sprinkle on sugar and add cream to taste. Whipped cream flavored with a liqueur of your choice can be substituted for the cream.

Berry Crumble

Crunchy! Fruity! Flavorful!

Berry layer
4 cups (1 l) mixed fresh berries (or frozen and thawed)
½ cup (125 ml) sugar
2 tablespoons (30 ml) cornstarch
½ teaspoon (2 ml) almond extract

Topping
½ cup (125 ml) quick oats
½ cup (125 ml) all-purpose flour
¼ cup (60 ml) packed brown sugar
¼ teaspoon (1 ml) salt
¼ cup (60 ml) butter or margarine
½ teaspoon (2 ml) cinnamon
¼ cup (60 ml) chopped pecans or walnuts

Preheat oven to 350°F (180°C).

Berry Layer: In a saucepan, combine berries with sugar and cornstarch. Heat and stir until mixture thickens to the consistency of jam. Stir in almond extract, and pour mixture into a medium non-stick baking pan.

Topping: Combine all topping ingredients with a fork until crumbly. Sprinkle over berry layer. Bake for 30 minutes or until filling is bubbly and topping is lightly browned.

Eat as is, or serve with whipped cream or vanilla ice cream.

Further Reading

Clark, Lewis J. 1998. (third edition). *Wild Flowers of the Pacific Northwest*. Harbour Publishing, Madeira Park, BC

Derig, Betty, B. & Margaret C. Fuller. 2001. *Wild Berries of the West*. Mountain Press Publishing Company, Missoula, MT

Johnson, Derek, L. Kershaw, A. MacKinnon & J. Pojar. 1995. *Plants of Western Boreal Forest & Aspen Parkland*. Lone Pine Publishing, Edmonton, AB

Kershaw, Linda, A. MacKinnon & J. Pojar. 1998. *Plants of the Rocky Mountains*. Lone Pine Publishing, Edmonton, AB

Lyle, Katie Letcher. 1994. *The Wild Berry Book: Romance, Recipes and Remedies*. Northword Press, Minnetonka, MN

Lyons, C.P. & Bill Merilees. 1995. *Trees, Shrubs & Flowers to Know in Washington & British Columbia*. Lone Pine Publishing, Edmonton, AB

MacKinnon, Andy, J. Pojar & R. Coupe. 1999. (second edition). *Plants of Northern British Columbia*. Lone Pine Publishing, Edmonton, AB

Pojar, Jim & A. MacKinnon. 1994. *Plants of Coastal British Columbia including Washington, Oregon and Alaska*. Lone Pine Publishing, Edmonton, AB

Pratt, Verna E. 1995. *Alaska's Wild Berries and Berry-like Fruits*. Alaskakrafts, Inc., Anchorage, AK

Turner, Nancy J. 1995. *Food Plants of Coastal First Peoples*. Royal British Columbia Museum Handbook. UBC Press, Vancouver, BC

Turner, Nancy J. 1997. *Food Plants of Interior First Peoples*. Royal British Columbia Museum Handbook. UBC Press, Vancouver, BC

Turner, Nancy J. & Adam F. Szczawinski. 1988. *Edible Wild Fruits and Nuts of Canada*. Canada's Edible Wild Plant Series, Vol. 3. National Museum of Natural Sciences & Fitzhenry & Whiteside, Markham, ON

Underhill, J.E. 1974. *Wild Berries of the Pacific Northwest: On the Bush, on the Table, in the Glass*. Hancock House Publishers, Saanichton, BC

Acknowledgements

I would like to thank several people who assisted with this project.

Susan Servos-Sept for her wonderful recipes.

Mary Schendlinger for her careful editing.

Jim Salt, who helped locate species for photography.

The skilled photographers who provided photos. Their names appear below.

Photo Credits

All photos by Duane Sept except the following:

Andy Fyon 16BR, 24TR, 62TL

Robert McCaw 23B, 52BR, 55B, 63BR

Susan Servos-Sept 95T

Cliff Wallis 15B, 24TL, 36BL, 71B

Cleve Wershler 55T

Index

Actaea
  arguta, 75
  eburnea, 75
  rubra, 75
Amelanchier alnifolia, 26
Aralia nudicaulis, 65
Arbutus menziesii, 70
arbutus, 70
Arctostaphylos
  alpina, 71
  columbiana, 71
  rubra, 71
  uva-ursi, 71
baked-apple, 36
baneberry, 75
  red, 75
  white, 75
bead ruby, 62
beadruby, Canada, 62
bearberry
  black alpine, 71
  common, 71
  red alpine, 71
Berberis aquifolium, 11
bilberry
  bog, 47
  dwarf, 47
bittercherry, 24
bittersweet, 78
  European, 78
  black berry, 65
blackberry
  Pacific, 34
  cutleaf, 33
  cut-leaved, 33
  evergreen, 33
  Himalayan, 32
  trailing, 34
blackcap, 31
blue bead, 60
blueberry
  Alaskan, 46
  bog, 47
  dwarf, 47
  oval-leaf, 46
  velvet-leaf, 47
bramble, five-leaf, 36

buffaloberry, 70
buffaloberry, Canada, 70
bunchberry, 43
bush-cranberry
  American, 55
  high, 54
  low, 54
  oval-leaf, 55
cascara, 67
cherry
  bird, 24, 64
  bitter, 24
  choke, 23
  fire, 24
  Pennsylvania, 24
  pin, 24
  red, 24
  wild, 23
chokecherry, 23
Clintonia uniflora, 60
clintonia, one-flowered, 60
cloudberry, 36
Comandra livida, 64
comandra, northern, 64
Convallaria majalis, 62
Cornus
  alba, 44
  canadensis, 43
  nuttallii, 44
  sericea, 44
  stolonifera, 44
  unalaschkensis, 43
crab apple, Pacific, 28
crabapple
  Oregon, 28
  Western, 28
crake berry, 65
cranberry
  Americanbush, 55
  bog, 52
cranberry
  small, 52
  wild, 52
Crataegus
  columbiana, 21
  douglasii, 21
  monogyna, 21

crowberry, 65
curlew berry, 65
currant
  prickly, 14
  bristly black, 14
  buffalo, 15
  golden, 15
  northern black, 14
  northern red, 16
  red flower, 15
  red swamp, 16
  red-flowering, 15
  squaw, 16
  sticky, 14
  wax 16
  wild red, 16
  winter, 15
deerberry, 62
devil's club, 66
dewberry, 34, 36
dewberry, western, 34
Disporum
  hookeri, 60
  trachycarpum, 60
dogberry, 64
dogwood
  Canada, 43
  dwarf, 43
  Pacific, 44
  red-osier, 44
doll's eyes, 75
Douglas berry, 34
Echinopanax horridum, 66
Elaeagnus commutata, 67
elder
  Pacific red, 57
  red, 57
  red-berried, 57
elderberry,
  black, 57
  blue 58
  red, 57
Empetrum nigrum, 65
Exobasidium vaccinii (fungus), 48
fairybells
  Hooker's, 60
  Oregon, 60
  rough-fruited, 60
Fragaria
  chiloensis, 19
  vesca, 19
  virginiana, 19

Gaultheria
  hispidula, 74
  shallon, 72
Geocaulon lividum, 64
gooseberry
  American mountain, 17
  black, 14
  black swamp, 14
  bristly wild, 17
  Canadian, 17
  northern, 17
  smooth, 17
  swamp, 14
grouseberry, 50
hawthorn
  black, 21
  common, 21
  Douglas, 21
  red, 21
highbush-cranberry, 54
holly, English, 76
honeysuckle
  bracketed, 69
  Utah, 68
  western trumpet, 69
huckleberry
  black, 49
  evergreen, 49
  false, 48
  littleleaf, 50
  red, 50
  western, 47
Ilex aquifolium, 76
Indian-plum, 64
juneberry, 26
juniper
  common, 8
  creeping, 9
  ground, 8
  rocky mountain, 9
Juniperus
  communis, 8
  horizontalis, 9
  scopulorum, 9
kinnikinnick, 71
laughing berries, 72
lily, bead, 60
lily-of-the-valley, 62
  false, 62
  wild, 62
lingonberry, 52

Lonicera
  ciliosa, 69
  involucrata, 69
  utahensis, 68
madrone, 70
mahogany, mountain, 74
Mahonia
  aquifolium, 11
  nervosa, 12
  repens, 12
Maianthemum
  canadense, 62
  dilatatum, 62
  racemosum, 61
  stellata, 61
  stellatum, 61
Malus fusca, 28
manzanita, hairy
mayflower, Canada, 62
Menziesia ferruginea, 48
moose berry, 54
mossberry, 65
mountain huckleberry, 49
mountain-ash
  European, 41
  Sitka, 41
  western, 41
nagoonberry, dwarf, 36
nightshade, woody, 78
Oemlaria cerasiformis, 64
Oplopanax horridus, 66
Oregon-grape
  creeping, 12
  dull, 12
  hollyleaf, 11
  tall, 11
Osmaronia cerasiformis, 64
oso-berry, 64
Oxycoccos
  microcarpus, 52
  quadripetalus, 52
  oxycoccos, 52
Pacific madrone, 70
Pembina, 55
plum, Indian, 64
poison oak, 77
poison-ivy, 77
Prosartes hookeri, 60
Prunus
  emarginata, 24
  pennsylvanica, 24
  virginiana, 23

pumpkin berry, 64
Pyrus fusca, 28
queen's cup, 60
raspberry
  American red, 30
  arctic, 36
  black, 31
  common red, 30
  creeping, 36
  dwarf, 36
  red, 30
  salmon, 35
  whitebark, 31
  wild red, 30
Rhamnus purshiana, 67
Rhus
  diversiloba, 77
  radicans, 77
Ribes
  aureum, 15
  cereum, 16
  hirtellum, 17
  hudsonianum, 14
  lacustre, 14
  oxyacanthoides, 17
  propinquum, 16
  sanguineum, 15
  setosum, 17
  triste, 16
  viscosissimum, 14
Rosa
  acicularis, 38
  gymnocarp, 39
  nutkana, 39
  pisocarpa, 39
  woodsii, 38
rose
  baldhip, 39
  clustered wild, 39
  common wild, 38
  Nootka, 39
  prickly wild, 38
  prickly, 38
  Wood's, 38
Rubus
  acaulis, 36
  arcticus, 36
  chamaemorus, 36
  discolor, 32
  idaeus, 30
  laciniatus, 33
  leucodermis, 31

parviflorus, 35
pedatus, 36
procerus, 32
pubescens, 36
spectabilis, 35
strigosus, 30
ursinus, 34
salal, 72
salmonberry, 35
Sambucus
  cerulea, 58
  melanocarpa, 57
  pubens, 57
  racemosa ssp. *pubens* var. *arborescens*, 57
  racemosa ssp. *pubens* var. *melanocarpa*, 57
sarsaparilla, wild, 65
saskatoon, 26
serviceberry, 26
  western, 26
Shepherdia canadensis, 70
silverberry, 67
skunk bush, 64
Smilacina
  racemosa, 61
  stellata, 61
snake berry, 75
snowberry
  common, 74
  creeping, 74
soapberry, 70
Solanum dulcamara, 78
Solomon-plume, 61
Solomon's-seal
  false, 61
  star-flowered, 61
  two-leaved, 62
  star-flowered false, 61
soopolallie, 70
Sorbus
  aucuparia, 41
  scopulina, 41
  sitchensis, 41
spikenard, false, 61
squashberry, 54
strawberry
  beach, 19
  coastal, 19
  mountain, 19
  smooth wild, 19
  wild, 19
  woodland, 19

Streptopus
  amplexifolius, 63
  roseus, 63
  streptopoides, 63
Svida sericea, 44
Symphoricarpos albus, 74
Taxus brevifolia, 74
thimbleberry, 35
thorn apple, western, 21
timberberry, 64
toad-flax, bastard, 64
Toxicodendron
  diversilobum, 77
  radicans, 77
twinberry
  black, 69
  red, 68
twistedstalk
  clasping, 63
  clasping-leaved, 63
  rosy, 63
  small, 63
Vaccinium
  alaskaense, 46
  caespitosum, 47
  membranaceum, 49
  microcarpus, 52
  myrtilloides, 47
  occidentale, 47
  ovalifolium, 46
  ovatum, 49
  oxycoccos, 52
  parvifolium, 50
  scoparium, 50
  uliginosum, 47
  vitis-idaea, 52
Viburnum
  edule, 54
  ellipticum, 55
  opulus, 55
  trilobum var. *americanum*, 55
watermelon berry, 63
waxberry, 74
whortleberry, 71
  grouse, 50
willow, red, 44
wolf-willow, 67
yew
  Pacific, 74
  western, 74

About the Author

Duane Sept is a biologist, freelance writer and professional photographer. His biological work has included research on various wildlife species and service as a park naturalist. His award-winning photographs have been published internationally, in displays and in books, magazines and other publications, for clients that include BBC Wildlife, Parks Canada, Nature Canada, National Wildlife Federation and World Wildlife Fund.

Today Duane brings a wealth of information to the public as an author, in much the same way he has inspired thousands of visitors to Canada's parks. His published books include *The Beachcomber's Guide to Seashore Life in the Pacific Northwest* (Harbour Publishing) and *Common Birds of British Columbia* (Calypso Publishing).